Studies in Computational Intelligence

Volume 1225

The series "Studies in Computational Intelligence" (SCI) publishes new developments and advances in the various areas of computational intelligence—quickly and with a high quality. The intent is to cover the theory, applications, and design methods of computational intelligence, as embedded in the fields of engineering, computer science, physics and life sciences, as well as the methodologies behind them. The series contains monographs, lecture notes and edited volumes in computational intelligence spanning the areas of neural networks, connectionist systems, genetic algorithms, evolutionary computation, artificial intelligence, cellular automata, self-organizing systems, soft computing, fuzzy systems, and hybrid intelligent systems. Of particular value to both the contributors and the readership are the short publication timeframe and the world-wide distribution, which enable both wide and rapid dissemination of research output.

Indexed by SCOPUS, DBLP, WTI AG (Switzerland), zbMATH, SCImago.

All books published in the series are submitted for consideration in Web of Science.

Thasayu Soisoonthorn · Herwig Unger

A Brain-Inspired Approach to Natural Language Processing

Springer

Thasayu Soisoonthorn
Bangkok, Thailand

Herwig Unger
Iserlohn, Germany

ISSN 1860-949X ISSN 1860-9503 (electronic)
Studies in Computational Intelligence
ISBN 978-3-032-00013-2 ISBN 978-3-032-00014-9 (eBook)
https://doi.org/10.1007/978-3-032-00014-9

This Springer imprint is published by the registered company Springer Nature Switzerland AG
The registered company address is: Gewerbestrasse 11, 6330 Cham, Switzerland

To my parents, Suthat and Sujitra Soisoonthorn, this book is dedicated with heartfelt gratitude for your unwavering love, support, and encouragement, which have been the cornerstone of everything I have achieved in my journey.

—Thasayu Soisoonthorn

To my wife, Dr. Panchalee Sukjit, who stands by my side always and supports me with all her strength.

—Herwig Unger

Preface

As human beings, our intelligence is what sets us apart from other animals. For thousands of years, we have been curious about how our minds work: how we think, learn, and solve complex problems. Our goal has not only been to understand this process but also to replicate it through artificial intelligence (AI).

One of the earliest approaches to building AI involved rule-based systems, which rely on human knowledge encoded as "if-then" statements. These systems are useful when speed and precision are important, and when large amounts of data are not required. However, rule-based systems cannot go beyond their programmed instructions. They lack the ability to adapt, make new inferences, or handle unexpected situations.

To improve upon these limitations, researchers developed knowledge-based systems that can reason with existing facts and make logical decisions. These systems were commonly used in expert systems to answer questions or assist in decision-making. Still, both rule-based and knowledge-based systems struggle with very large or complex problems, especially when there are many possible solutions.

In cases where it is difficult to define clear rules or predict every possible situation, we often rely on data instead. This leads us to machine learning (ML), which allows computers to learn patterns from data and improve over time without being explicitly programmed for each task. Popular ML techniques include ensemble models such as Random Forests and Gradient Boosting, k-Nearest Neighbors (kNN), Support Vector Machines (SVM), and Deep Learning (DL), which has driven many recent breakthroughs in AI.

However, traditional ML methods are mostly designed for one-time decisions in what are called episodic environments. They tend to struggle with tasks that unfold over time. When the order and timing of information matter, such as in language understanding or behavioral analysis, we need models that can interpret sequences. To address this, researchers introduced time-series methods like ARIMA and neural models such as Recurrent Neural Networks (RNNs), Long Short-Term Memory networks (LSTMs), and Gated Recurrent Units (GRUs). These models enable machines to process and learn from data that changes over time, though they face limitations in handling longrange dependencies and require sequential processing that can be computationally slow. The introduction of the Transformer architecture revolutionized sequence modeling by replacing sequential processing with parallel attention mechanisms. This breakthrough led to the development of Large Language Models (LLMs) like GPT, which have achieved remarkable capabilities through self-attention mechanisms.

Even with these advanced approaches, limitations persist. Most ML models learn passively by analyzing large, static datasets. But what happens when data is scarce or when learning must occur through interaction? This is where Reinforcement Learning (RL) becomes essential. RL allows agents to learn through trial and error by actively exploring their environment and discovering which actions lead to better outcomes. This process mirrors how humans and animals learn from experience.

While modern techniques like deep learning have achieved remarkable results, yet they also come with significant trade-offs. These models require vast computational resources, consume substantial energy, and often depend on enormous datasets. They function as black boxes, making it difficult to interpret their decision-making processes. Moreover, they are sensitive to small input variations, prone to errors, and lack true reasoning or understanding. Even powerful systems like Large Language Models (LLMs), which are based on the Transformer architecture, face challenges in accuracy, adaptability, and common sense. While they can generate fluent text and understand context, they may hallucinate facts and cannot continuously learn from real-world feedback.

Although AI and its technologies have evolved significantly and produced impressive outcomes, they have not yet reached the level of human intelligence in several key areas. Since intelligence originates in the human brain, it is reasonable to look to the brain for guidance on how AI systems should be designed.

This book takes readers on a thoughtful journey into how neuroscience can inspire the development of AI systems. Chapters 1 and 2 establish the biological foundation by examining the brain's structure, evolution, and functions, with a focus on the neocortex and its role in learning and prediction. These chapters introduce Jeff Hawkins' theories, which link neuroscience to computational models. Chapter 3 builds on these ideas by surveying traditional and modern AI methods and highlighting their limitations. This contrast underscores the potential of brain-inspired approaches to address existing gaps. Building on this foundation, Chap. 4 presents Sparse Distributed Representations (SDRs), a brain-like encoding mechanism.

The focus then shifts from theory to practice. Chapters 5–7 demonstrate how these concepts can be applied to real-world NLP tasks, such as spelling correction and Thai word segmentation. These chapters validate the model's adaptability, noise tolerance, and low computational cost.

Finally, Chap. 8 concludes with a reflection on the book's overall contributions and offers suggestions for future exploration. The book provides a cohesive journey from brain biology to NLP, offering fresh insights at the intersection of neuroscience and intelligent systems.

The purpose of this book is to spark curiosity, promote hands-on experimentation, and support ongoing progress in the development of intelligent systems. The method introduced here is not intended to replace current approaches or to claim that it is the path toward building a general-purpose AI. Instead, it explores a key question: Can knowledge from the human brain be adapted, integrated, or reimagined to enhance the way AI solves complex problems?

The approach presented in this book is intentionally kept simple and easy to interpret. It draws inspiration from biological processes in the brain while aiming to deliver practical and effective results. As discussed in Sect. 3.9, there is strong potential for improving current AI techniques by applying insights from cognitive neuroscience.

While refining existing AI models may lead to progress, it is uncertain whether those paths alone can achieve true intelligence. In contrast, designing AI systems based on how the human brain works may offer a promising alternative that could lead to more flexible, adaptive, and human-like behavior.

With this perspective, the book encourages advanced learners to reflect, experiment, and think independently. We hope readers enjoy engaging with the material and are inspired to contribute new ideas. Feedback and suggestions are warmly welcomed, as they help shape the future of intelligent system development.

Bangkok, Thailand
Iserlohn, Germany
May 2025

Thasayu Soisoonthorn
Herwig Unger

Competing Interests The authors have no competing interests to declare that are relevant to the content of this manuscript.

Contents

Chapter 1
The Human Brain

To develop intelligent systems that reflect human abilities, it is crucial to first understand the brain that inspires them. This chapter examines the brain's evolution, structure, and learning mechanisms, offering key insights into how biological intelligence functions. It provides the foundational knowledge needed to appreciate the principles underlying human cognition.

The chapter begins with a comparison of the brain's performance and energy efficiency relative to artificial systems (Sect. 1.1). This is followed by an overview of its evolutionary development (Sect. 1.2), tracing the progression from primitive nerve nets to the complex mammalian brain. It then explores the brain's functional structure (Sect. 1.3), highlighting regions such as the cerebellum, hippocampus, thalamus, limbic system, and especially the neocortex (Sect. 1.4), which supports higher-level functions such as reasoning and voluntary movement. The neocortex's global structure, cortical areas, and reward system are also discussed.

Subsequent sections cover neuronal processing (Sect. 1.5), including neuron structure, action potentials, synaptic plasticity, and memory formation, along with insights into human forgetting and learning patterns. The chapter then explains how the brain filters and manages sensory information (Sect. 1.6), showing how different types of memory and senses work together to keep thinking efficient. A summary of the main ideas from the chapter is provided in Sect. 1.7.

1.1 Overview and Comparison of Performance

The human brain is still the most powerful information processing and communication system on Earth. By understanding how it stores information, learns, and recognizes objects and processes, we can develop computers and other artificial systems that mimic or are inspired by these abilities (see also [1]).

T. Soisoonthorn and H. Unger, *A Brain-Inspired Approach to Natural Language Processing*, Studies in Computational Intelligence 1225,
https://doi.org/10.1007/978-3-032-00014-9_1

The average adult human brain weighs around 1,400 g, which is about 2–5% of the total body weight. It contains roughly 86 billion neurons, which are responsible for processing and transmitting information. Supporting these neurons are nearly the same number of glial cells, which help by providing structure, insulation, and transporting nutrients and waste. Without oxygen and nutrients, the brain stops functioning, and its stored information is permanently lost within 6–10 min.

Each neuron is typically connected to about 1,000 other neurons, forming a vast network that powers human creativity and thinking. This complex system runs on about 20 w of energy, making up 20–50% of the total energy used by adults and newborns, respectively.

Scientists have already managed to replicate some parts of the brain, especially neurons and their networks, which will be explored further in the next section. Artificial neurons can work up to 100,000 times faster than natural ones. However, the brain's true power comes from its ability to process information in parallel. While artificial systems have made great strides, they still only match the brain's abilities in specific areas.

Examples of Achievements and Comparisons

- **Intel's Hala Point (2025)**: The Hala Point chip [2] features 1.15 billion artificial neurons and 128 billion synapses, marking a significant milestone in neuromorphic computing. Its sophisticated architecture demonstrates exceptional scalability and energy efficiency, helping to bridge key performance gaps between artificial and biological neural systems. Although its neural capacity, measured by the number of neurons and synapses, is roughly comparable to that of small mammals, such as an owl, the chip still falls far short of replicating the complexity and adaptability of the human brain.
- **Computational Power**: Comparing artificial systems to the human brain's computational capacity is complex. While systems like IBM's TrueNorth [3] and Intel's Hala Point have impressive computational capabilities, directly equating them to the brain's estimated 10^{13} analog operations per second [4] is challenging due to fundamental differences in architecture and processing methods. The brain's remarkable efficiency comes from its massively parallel processing, where countless operations happen simultaneously across vast neural networks.
- **Energy Efficiency**: The energy consumption of these artificial systems highlights a stark difference. Intel's Loihi [5] neuromorphic chips are up to 16 times more energy-efficient than conventional AI hardware, consuming significantly less power for deep learning tasks. Similarly, Intel's Hala Point chip achieves 15 TOPS/W efficiency, consuming around 2.1 kW for high-performance workloads. While these advancements mark substantial improvements over traditional systems like Blue Gene/L (1.2 MW) [6] and Tera-Scale (62 W) [7], they remain far less efficient than the human brain, which performs approximately operations per second using only 20 W [8].
- **Structural Complexity**: Despite advances, artificial systems still lack the adaptive neuroplasticity of the brain, relying on fixed network structures rather than

dynamic rewiring. This limitation makes them less resilient and more vulnerable to disruption.
- **Robustness and Fault Tolerance**: Biological neurons exhibit remarkable fault tolerance, compensating effectively for damage or loss. In contrast, artificial systems typically experience severe disruptions from partial failures, lacking this natural resilience.
- **Quantum Computing**: Quantum computers leverage qubits in superposition to efficiently tackle tasks such as cryptography, complex simulations, and optimization. Google's Sycamore processor famously completed a computation in 200 s that would take 10,000 years on a classical supercomputer [9]. The integration of quantum computing with AI has led to the emergence of Quantum Artificial Intelligence (Quantum AI), enhancing AI capabilities through techniques like Quantum Neural Networks (QNNs) [10] and Quantum Support Vector Machines (QSVMs) [11], supported by platforms such as TensorFlow Quantum [12]. Microsoft's Majorana 1 quantum processor [13] utilizes topological qubits and novel materials known as topoconductors, offering improved stability and scalability. While quantum systems promise potential energy efficiency gains [14], they typically require extreme cooling [15]. Despite rapid progress, quantum computing remains in its early stages, limited by low qubit counts and high error rates, which continue to hinder practical applications.

The brain's extraordinary efficiency and performance are rooted in its evolutionary development over millions of years. Beginning with simple sensory cells, countless variations and mutations have led to the survival and further evolution of only the most effective structures. This evolutionary process has shaped the human brain into the unparalleled system it is today. The following sections will provide a brief overview of this developmental journey.

1.2 Outline of The Brains Evolution

Less than a billion years after the formation of the Earth 4.5 billion years ago, single-celled prokaryotic life emerged, comprising cells that lack a cell nucleus. It took an additional billion years for eukaryotic life, consisting of cells with a nucleus, to appear. From there, it took over another billion years for multicellular life to develop from eukaryotic cells. While some multicellular organisms initially relied on flagella for movement, it quickly became clear that this method alone was not efficient for swift and coordinated movements. As a result, networks of specialized cells evolved, enabling rapid electrical signaling throughout neuron networks, which synchronized external structures for movement. One of the earliest and simplest nervous systems can be observed in jellyfish, which have an undifferentiated nerve net that is not organized into distinct parts like a brain or spinal cord. In this nerve net, dedicated nerve cells are interconnected, communicate effectively, and send targeted signals to muscle cells, enabling coordinated contractions and improved movement control.

Neurons in these primitive nervous systems evolved distinctive features allowing the processing and communication of vital environmental information. This evolution significantly enhanced an organism's chances of survival by enabling the detection of food sources, temperature variations, predators, and toxins, thus facilitating adaptive movement within their environment.

Approximately 500 million years ago, invertebrates, including insects, began to inhabit the land, consuming the plants that had been growing for millions of years. This transition marked a critical phase in the evolution of life on Earth, introducing more complex interactions between organisms and their environments. Among these invertebrates, insects and arthropods began developing the earliest simple, centralized nervous systems known as ganglion or ladder nerve systems [16]. In these systems, nerve nodes, typically arranged in pairs, interconnect in a ladder-like structure (see Fig. 1.1).

Cockroaches exemplify ganglion nerve systems, possessing two brain-like command centers located in the head, along with additional nerve nodes distributed throughout the body. These nodes often connect directly to muscles through motor nerves, enabling rapid and direct responses to environmental stimuli without solely relying on the brain as the central control unit for the entire nervous system and body.

The simplicity of such nerve systems has allowed researchers, including Erickson et al. [17–19], to study cockroach movements by applying electrical stimuli to specific nerve nodes (see Fig. 1.2). A small computer transmits electrical impulses via wires to these nerve nodes, enabling precise control over the cockroach's direction, such as turning left, turning right, or moving straight ahead. When combined

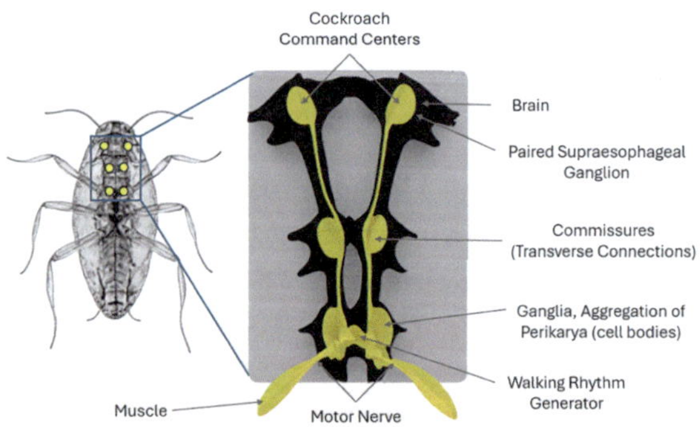

Fig. 1.1 Structure of the Cockroach Nervous System

Fig. 1.2 Cyber Cockroaches [18]

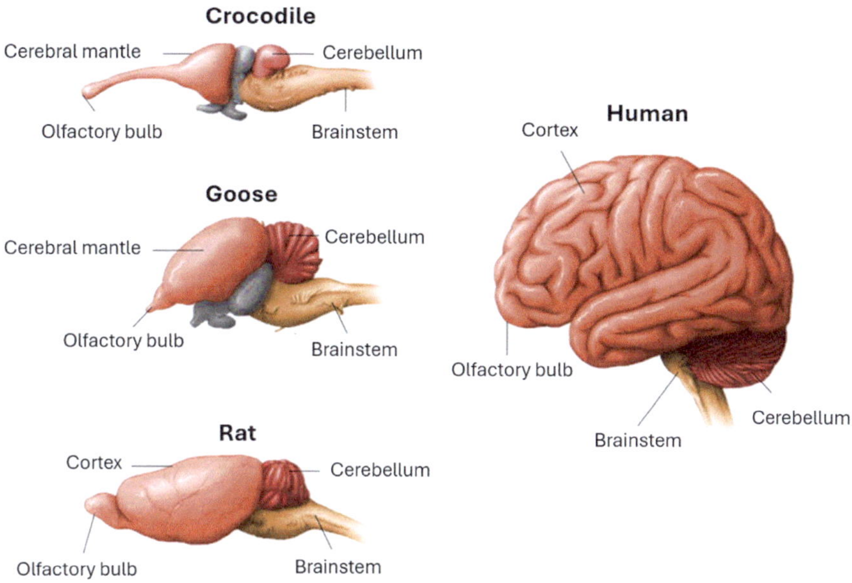

Fig. 1.3 The Development of the Mammalian Brain

with additional sensors such as miniature cameras, this technique offers practical applications, particularly in fields like disaster management.

As evolution progressed, some vertebrate lungfishes were forced to travel onto land after the water dried up, struggling to reach larger pools. These species gradually adapted to terrestrial life, evolving into amphibians, then reptiles, and ultimately into mammals. Throughout this evolutionary pathway, there was a clear trend toward the increased concentration and centralization of nerve cells. This culminated in the development of a true centralized brain, first notably observed in mammals [20] (see Fig. 1.3), and ultimately leading to the emergence of humans less than a million years ago.

The brain is a crucial organ that serves as the central nervous system in mammals, non-mammalian vertebrates, and invertebrates, which are animals without backbones. In simple animals, the brain acts as a switchboard, receiving signals from sensory organs and transmitting information to muscles to carry out behaviors such as walking, flying, finding food, and mating. In humans, the brain governs various essential functions such as the ability to speak, think, hear, see, feel, remember, walk, and plan. At the basic level, the working unit of the brain is a neural cell, a specialized cell designed to transmit information to other nerve cells, gland cells, or muscles. The brain's ability to function stems from the structural and functional characteristics of interconnected neurons. The brain has evolved over time, leading to the following developments.

- **Invertebrates**, such as insects, mollusks, and sponges, are the oldest known multicellular animals in terms of evolution. However, they have limited cognitive capacity due to the distributed organization of their neurons and the small size of their nervous systems. For instance, roundworms possess a nervous system consisting of roughly 300 neural cells. The sensory system in the head end of these organisms detects environmental stimuli and conveys signals to the brain, which in turn sends messages to muscles to elicit a response to the stimulus.
- **Non-mammalian vertebrates**, such as reptiles (e.g., alligators and salamanders), amphibians (e.g., frogs), birds, and fish, have relatively centralized brains with many structures that resemble those found in mammals. However, they lack a key feature of mammalian brains: the neocortex. Instead, their brains consist of specialized regions that handle sensory processing and behavior control, but these brains are generally smaller and less complex.
  The organization of the brain is often described in terms of a three-level hierarchy. At the lowest level is the basal ganglia, which governs instinctive behaviors such as fight-or-flight responses, defending territory, mating pursuits, and hunger, functions commonly seen in the so-called reptilian brain (e.g., in lizards or snakes). Above this is the limbic system, which includes the hypothalamus, amygdala, hippocampus, septum, and cingulate cortex. This system supports emotions and enables more complex behaviors like social bonding and parental care. The highest level of brain organization is the neocortex, responsible for reasoning and advanced cognition, but this structure is unique to mammals and is absent in non-mammalian vertebrates.
- **Mammals** came into existence during the middle of the dinosaur era and emerged as the dominant land animals after the extinction of the dinosaurs roughly 65 million years ago. Compared to non-mammalian vertebrates, mammals share a greater similarity in brain structure, with the key difference being the presence of the neocortex. The neocortex is called "neo" because it is a relatively recent evolution in mammals, situated on top of old parts of the brain. A large neocortex, which reaches its peak in primates, enables intelligence, complex social behavior, consciousness, and awareness. The neocortex provides the advantage of rapid learning when environments change quickly. While all plants and animals can adapt to changing environments over time, they must rely on genetic evolution without a neocortex, which can take many generations or thousands of years.

1.3 Functional Structuring

The brain consists of multiple functional areas that have evolved over time, from invertebrates to mammals. Although neuroscientists do not yet fully understand how the brain works, its functions can be broadly inferred through research and studies of brain injuries. This chapter introduces key concepts by highlighting various brain regions and their crucial roles in cognitive functions.

- **Cerebellum**, often referred to as the "little brain," is one of the oldest and most conserved structures in the vertebrate brain. It contains more neurons than the rest of the brain combined, including the neocortex, and accounts for roughly 80% of total neurons. The cerebellum plays a critical role in motor control, coordination, precision, and timing. Dysfunction in this area often results in noticeable motor deficits. In addition to its motor functions, the cerebellum also contributes to certain cognitive and emotional processes.
- **Thalamus**, located near the center of the brain, acts as a major relay hub for sensory and motor information. All sensory signals, except for smell, are routed through the thalamus before being processed in the neocortex. The olfactory system is unique because it bypasses the thalamus and connects directly to the neocortex. In addition to its sensory relay function, the thalamus plays a key role in regulating sleep, alertness, wakefulness, and consciousness. It serves as the brain's central switchboard, coordinating incoming and outgoing neural signals.
- **Limbic system** is an evolutionarily older network of brain structures that integrates memory, emotion, and behavior. It includes the hippocampus, amygdala, cingulate cortex, hypothalamus, and other associated areas. This system is central to processing emotions and linking them with memories and motivations. Even vertebrates such as lizards, which lack a neocortex, demonstrate the capacity for learning and memory through the limbic system. For instance, if a lizard loses a fight and later encounters the same opponent, it may exhibit fear. Similarly, returning to a familiar feeding ground can trigger hunger-related behaviors. These emotionally influenced memories help animals adapt behaviorally to their environment.
- **Hippocampus**, a core component of the limbic system, serves three primary functions: spatial cognition, response inhibition, and episodic memory. It is widely recognized for its role in forming new memories about experienced events. Episodic memory involves detecting novel events, places, and stimuli and temporarily storing them. The hippocampus receives input from across the neocortex and sends projections back to those same regions. It can replay sequences of events, reactivating the specific neocortical areas involved in the original experience. This memory replay, which often occurs during REM (Rapid Eye Movement) sleep, facilitates the consolidation of short-term memories into long-term storage. Remarkably, the hippocampus allows even early-evolved vertebrates such as birds and reptiles to learn from experience in the absence of a neocortex.
- **Amygdala**, another major structure within the limbic system, works closely with the hippocampus but specializes in processing emotionally significant memories. It detects emotionally salient stimuli, particularly those related to fear, threat, or survival, and generates instinctive autonomic responses. For example, the amygdala may trigger fear in response to spiders, snakes, or expressions of distress in others. These responses are typically unconscious and rapid, serving as a crucial mechanism for survival.
- **Neocortex** is the most recently evolved brain region, unique to mammals, and is responsible for higher cognitive functions such as perception, reasoning, problem-solving, and language. Organized into six layers, it integrates sensory inputs, facilitates abstract thinking, and enables voluntary motor control, referring to

the conscious planning and execution of movements like walking, grasping, or speaking. Unlike the limbic system, which governs emotions, the neocortex supports decision-making and conscious awareness, distinguishing mammals by their intelligence and adaptability.

Building upon this functional overview, the neocortex stands out as the most evolutionarily advanced region, integrating inputs from subcortical structures like the thalamus, hippocampus, and amygdala to enable higher-order cognitive functions. Its unique structure and capabilities are further explored in the following section.

1.4 Neocortex

1.4.1 Global Structure

Intelligent and adaptive behavior in mammals is primarily associated with the neocortex, which sits atop the evolutionarily older parts of the brain. Intelligence is closely linked to several anatomical features: the size of the neocortex, the brain-to-body mass ratio, and especially the size of the frontal lobe. The prefrontal cortex, in particular, plays a crucial role in abstract reasoning and high-level goal planning. This region is most developed in humans.

Two attributes often associated with human intelligence are episodic memory and language. Language appears to be unique to humans and is defined as the use of symbolic sequences structured by complex grammar. Episodic memory refers to the ability to recall specific events within a temporal and contextual framework. For example, one might say, "I can't remember where I put the keys". This contrasts with semantic memory, which involves general knowledge and associations, such as "I know where I am and can recognize objects around me". While other brain regions also play important roles in cognition, this book focuses on the neocortex, as it is a defining feature of human intelligence and underlies our most advanced cognitive abilities.

The neocortex is the largest part of the brain and is visible from above or the side. It is approximately a 1.5 square foot sheet of cells, which is slightly folded to fit inside the head, and accounts for 80% of the brain's weight. Homo sapiens developed a large forehead to accommodate the neocortex. The frontal lobe, in particular, helps us deal with abstract patterns associated with high-level concepts.

The functional structure of the neocortex is that of a folded sheet, as displayed in Fig. 1.4. These folds allow the surface area of the neocortex to be greatly increased and fit into a human head. The outermost 3 to 4 millimeters of this sheet form what is called *gray matter*, which contains many neural cell bodies that are darker than their axons. Underneath the gray matter are axons, called *white matter*, which consist almost entirely of the myelinated axons of neurons going to and from the neocortex to other areas.

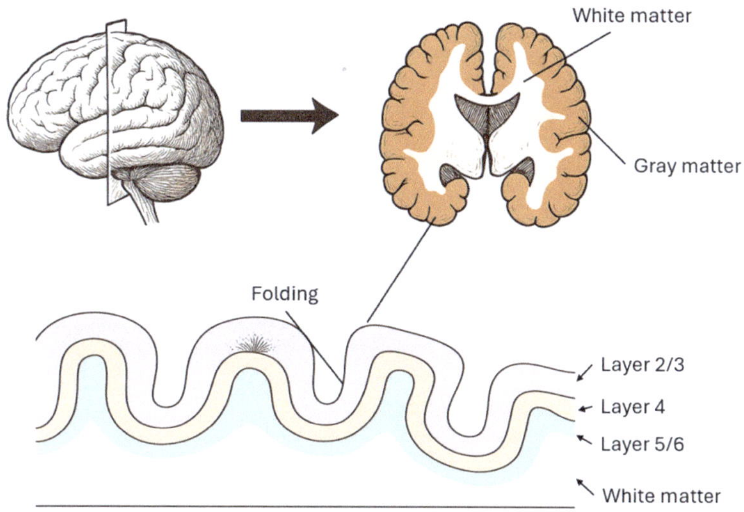

Fig. 1.4 White matter and gray matter in the folding of the neocortex

In addition to its overall role in enabling higher intelligence, the neocortex is organized into several specialized regions known as cortical areas, each responsible for specific cognitive and sensory functions. These areas are distributed across the four major lobes of the brain: the frontal, temporal, occipital, and parietal lobes (Fig. 1.5). For instance, Broca's area, located in the left frontal lobe, is crucial for language production, while Wernicke's area in the left temporal lobe supports language comprehension. The somatosensory cortex in the parietal lobe processes tactile and proprioceptive input, and the visual cortex in the occipital lobe is specialized for visual perception.

Beyond these sensory and motor regions, the neocortex also contains association areas that integrate information across modalities and support complex cognitive functions such as attention, memory, language, and decision-making. These interconnected cortical regions form the foundation of our capacity for abstract reasoning, conscious thought, and intentional behavior. To better understand how these specialized functions are distributed across the brain, the following subsections examine each of the four major lobes in detail.

1.4.1.1 Frontal Lobe

The frontal lobe, located at the front of the brain, is essential for a range of high-level cognitive and motor functions. It plays a central role in executive processes such as decision-making, planning, reasoning, and the initiation of voluntary movement. Within this lobe, the prefrontal cortex is particularly crucial, as it supports complex cognitive abilities including problem-solving, attention regulation, and impulse con-

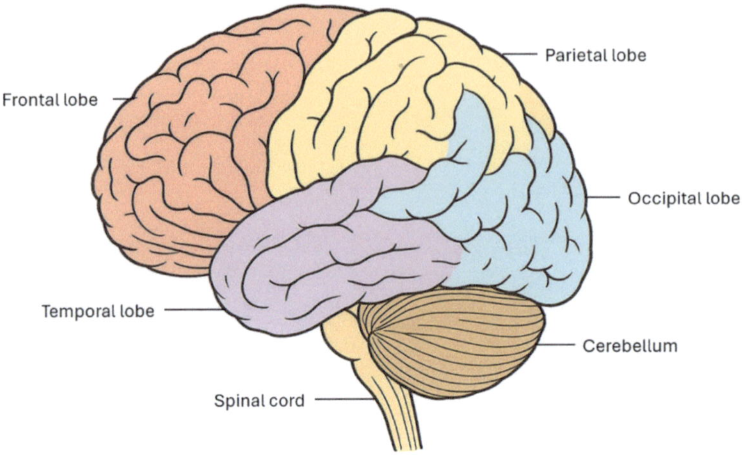

Fig. 1.5 Neocortex areas

trol. It also contributes significantly to personality development and social behavior, making it a key element in defining human individuality.

Motor function is another vital responsibility of the frontal lobe. The primary motor cortex, found in the precentral gyrus, directs voluntary movements by transmitting signals to specific muscles throughout the body. Nearby motor regions help refine these actions, coordinating complex behaviors such as speech. Among them, Broca's area, typically located in the left (dominant) hemisphere, is critical for speech production and the construction of grammatically coherent sentences.

The frontal lobe also has an important role in emotional and social regulation through its connections with the limbic system. It helps manage emotional reactions and interpret social cues, facilitating appropriate behavior in diverse situations. Injury to this region can lead to significant shifts in personality, impaired judgment, and diminished self-control. Overall, the frontal lobe's wide-ranging responsibilities make it essential for both cognitive performance and emotional well-being.

1.4.1.2 Temporal Lobe

The temporal lobe, situated on the sides of the brain beneath the lateral fissure, plays a vital role in auditory processing, memory formation, and language comprehension. It contains the primary auditory cortex, which interprets sound signals from the ears, enabling us to perceive and distinguish a wide range of sounds, including speech. This area is key to identifying pitch, rhythm, and tone, which are fundamental elements of auditory perception.

Memory is another major function of the temporal lobe. Deep within it lies the hippocampus, a critical structure for forming and retrieving long-term memories.

By linking new sensory information to past experiences, the hippocampus supports learning and helps maintain a continuous sense of personal identity.

Language comprehension is also mediated by the temporal lobe through Wernicke's area, typically located in the dominant hemisphere. This region allows us to interpret both spoken and written language, making it essential for meaningful communication. Additionally, the temporal lobe's visual association areas, including the fusiform gyrus, contribute to the recognition of faces and objects. Through its integration of auditory, visual, and memory-related functions, the temporal lobe serves as a core hub for understanding, communication, and recognition.

1.4.1.3 Occipital Lobe

The occipital lobe, located at the rear of the brain, serves as the primary center for visual processing. At its core lies the primary visual cortex, which receives raw visual input from the retina via the optic nerves. This region is responsible for processing basic visual features such as brightness, color, shape, and motion.

Following this initial stage, adjacent visual association areas within the occipital lobe perform higher-level interpretation. These areas enable us to recognize objects, perceive depth and movement, and understand spatial relationships. For example, they allow us to distinguish a face from its background or identify an object based on its outline.

The occipital lobe operates in close coordination with other brain regions, such as the parietal lobe for spatial awareness and the temporal lobe for object recognition, to construct a unified and meaningful visual experience. Damage to the occipital lobe can result in significant visual impairments, including partial blindness or visual agnosia, which is the inability to recognize familiar objects or faces.

1.4.1.4 Parietal Lobe

The parietal lobe, positioned between the frontal and occipital lobes, is chiefly responsible for integrating sensory information from multiple sources. Its most prominent structure, the primary somatosensory cortex located on the postcentral gyrus, processes tactile sensations such as touch, pressure, pain, and temperature. This area enables us to perceive physical stimuli and interpret their spatial distribution across the body.

Beyond basic sensation, the parietal lobe plays a critical role in spatial awareness and navigation. By combining visual and sensory input, it helps us orient ourselves within our environment, judge distances, and understand the position of our body parts in space. This function is known as proprioception.

The parietal lobe also supports advanced cognitive functions, including numerical reasoning and the manipulation of objects. Through its somatosensory association areas, it interprets and synthesizes sensory data to enhance our understanding of object characteristics like size, shape, and texture, whether perceived by touch or sight. These integrative functions are essential for coordinating perception with purposeful movement and problem-solving.

1.4.2 Reward System

The reward system is vital for motivating behavior toward stimuli such as food, sex, and alcohol. It also plays an important role in helping individuals avoid pain and conflict. Connected to brain regions responsible for behavior and memory, the reward system links actions with pleasure, encouraging the repetition of those behaviors.

The primary source of dopamine in the neocortex is the ventral tegmental area (VTA), where neurons release dopamine. This release generates a sense of wanting or motivation, though it does not necessarily produce pleasure. The feeling of pleasure arises from opioids in the brain that create a sense of happiness. Dopamine and opioids work together, and the amount of dopamine released corresponds to the potential for experiencing pleasure. When an activity is pleasurable, the brain forms a memory of that experience and expects a similar level of satisfaction in the future. If the actual experience falls short of or exceeds that expectation, the amount of dopamine released changes accordingly during future encounters. This variation influences how strongly the stimulus is perceived as rewarding. The mechanisms of this reward system form a core principle in reinforcement learning, as discussed in Sect. 3.8.

After understanding the functional structure of the brain, including the neocortex and the reward system, it is important to explore the neural foundations that support these functions. Since all cognitive and behavioral processes arise from neural activity, understanding how neurons transmit, process, and communicate information is essential. The next section will examine the fundamental properties of neurons and the mechanisms that enable them to form complex neural circuits.

1.5 Neuronal Information Processing

1.5.1 Neurons

Neurons are the fundamental units of the brain responsible for processing information. They consist of three main parts, each with a specific role in how the brain receives, integrates, and sends signals as shown in Fig. 1.6.

- **Dendrites**: Tree-like structures that receive input signals from sensory organs or other neurons. A single neuron can receive signals from thousands, and in some cases up to 100,000 [21], different sources, making dendrites essential for collecting and integrating diverse information

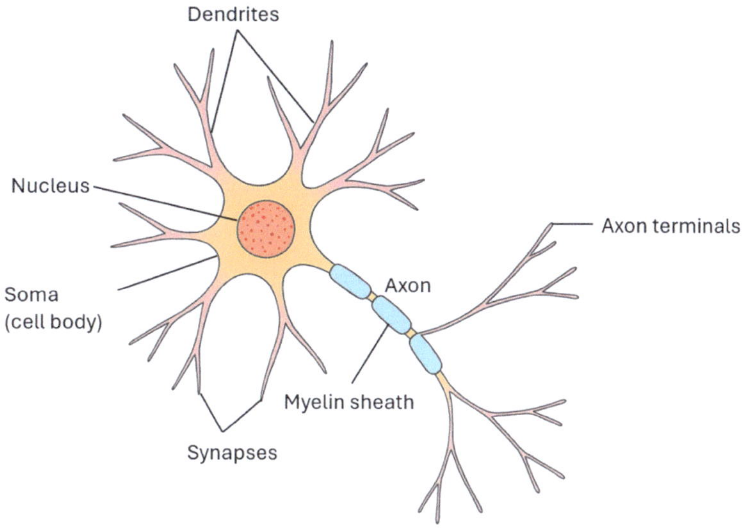

Fig. 1.6 The structure of a neural cell

- **Soma (Cell Body)**: This is the processing center of the neuron. It sums all incoming excitatory and inhibitory signals and determines whether to trigger an action potential. The soma typically ranges in size from 5 to 100 micrometers and contains the axon hillock, the site where action potentials are initiated.
- **Axon**: A long projection that transmits the output signal to other neurons or muscles once the neuron is activated. Axons vary in thickness (0.05 to 20 μm) and can be as short as 1 micrometer or as long as 1 m in humans.

Neurons exhibit diverse structures and functions, with their structural variations illustrated in Fig. 1.7. They are generally classified based on the number of their extensions as follows:

- **Unipolar neurons**: These have a single axon and no dendrites. They are typically sensory neurons, such as those found in the retina.
- **Bipolar neurons**: With one axon and one dendrite, these neurons are involved in sensory pathways, including taste, touch, hearing, and balance.
- **Multipolar neurons**: The most common type, these have multiple dendrites and one axon. They are found in motor and interneurons, such as those in the spinal cord.
- **Pseudounipolar neurons**: These possess a direct connection between the dendrite and axon, allowing rapid signal transmission. They are found in sensory neurons responsible for fast reflexes.

With the structural components and classifications of neurons established, the next focus is on the fundamental mechanism of neuronal communication. This process

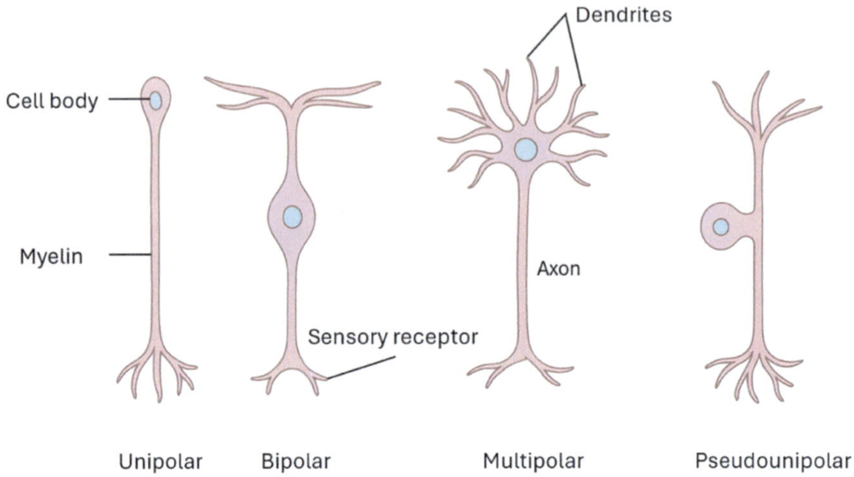

Fig. 1.7 Different Types of Neurons in the Brain

is driven by rapid electrical impulses, known as action potentials, which serve as the primary means of transmitting information across the brain's complex signaling networks.

1.5.2 Action Potentials

Neurons are the fundamental units of brain function, responsible for transmitting information through both electrical and chemical signals. This process begins when a neuron receives enough stimulation to generate an *action potential*, which is a rapid electrical impulse typically measuring around 70 millivolts. The signal travels from the cell body along a long fiber called the *axon* toward the synaptic terminal at the end of the neuron.

To increase the speed of this transmission, many axons are covered with a protective insulating layer called the *myelin sheath*. This sheath enables the signal to jump between gaps known as nodes of Ranvier, significantly accelerating signal conduction along the axon.

When the action potential reaches the end of the axon, it arrives at a specialized junction called the *synapse*, where the neuron communicates with another cell. There are two main types of synapses:

- **Electrical synapses**: allow direct electrical current to flow from one neuron to another via small channels called gap junctions. These synapses are extremely fast and enable synchronized activity, but are less common in the human brain.
- **Chemical synapses**: which are more prevalent, work through the release of *neurotransmitters*, or chemical messengers stored in small sacs called synaptic vesicles.

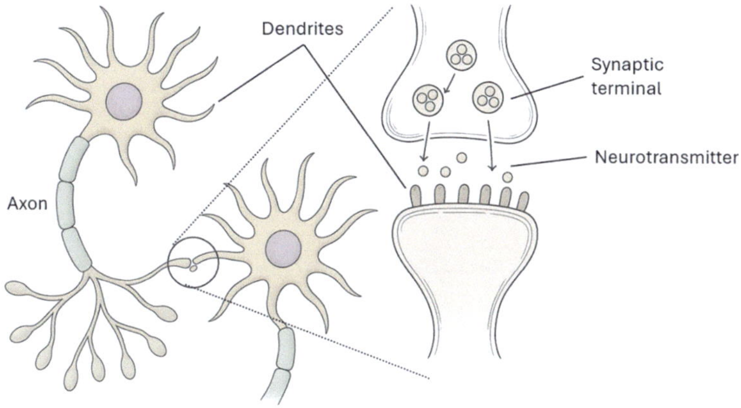

Fig. 1.8 Illustration of neurotransmitter release and synaptic transmission

When the action potential arrives, these neurotransmitters are released into the *synaptic cleft* and bind to receptors on the postsynaptic neuron.

Depending on the type of neurotransmitter and receptor involved, the signal can either *excite* the postsynaptic neuron by making it more likely to generate an action potential, or *inhibit* it by reducing the likelihood of firing. This dynamic balance between excitation and inhibition enables the nervous system to regulate neural activity with precision, preventing both overstimulation and complete shutdown of communication.

These electrical and chemical processes work together to enable neurons to transmit information rapidly and accurately across the brain and body. Figure 1.8 illustrates the key steps involved in this communication process.

1.5.3 Synaptic Plasticity: Learning and Memory Formation

While action potentials facilitate the immediate transfer of information, the brain must also adapt and store information over time. This ability is governed by synaptic plasticity, which allows synapses to strengthen or weaken based on experience. Synaptic connections are not static; they change based on activity, enabling learning and memory formation. The strength of a synapse can be increased or decreased through mechanisms such as Long-Term Potentiation (LTP) and Long-Term Depression (LTD):

- **LTP** is the process by which frequently used synapses become stronger, increasing the efficiency of communication between neurons. This enhanced connectivity makes it easier for signals to pass through the network, playing a crucial role in learning and memory. LTP follows the principle that frequent use of a neural path-

way reinforces its connections, ensuring that important information is retained. LTP exemplifies the principle of “what fires together, wires together”, proposed by Donald Hebb [22]. When two neurons are activated simultaneously, their connection is strengthened, forming the basis of Hebbian learning.
- **LTD**, on the other hand, weakens synapses that are used less frequently. This selective reduction in connectivity helps refine neural networks, eliminating redundant or outdated pathways to optimize brain function.

These processes are essential for cognitive flexibility, allowing the brain to adapt to new experiences, reinforce learned behaviors, and discard irrelevant information. Without synaptic plasticity, the nervous system would lack the ability to evolve with experience, making learning and memory formation impossible.

The dendrites, axons, and synapses work in unison to process and store information. Dendrites receive thousands of inputs from other neurons, and their synaptic connections are key to determining the neuron’s response. A neuron’s threshold for firing is influenced by the strength and combination of its inputs. This ensures that only significant patterns of activity lead to an action potential, conserving energy and optimizing the brain’s functionality.

The human brain contains approximately 86 billion neurons, forming trillions of synaptic connections. Each neuron’s ability to adjust synaptic weights, which represent the relative importance of inputs, underpins the brain’s capacity to recognize patterns and adapt to new information. This synaptic plasticity is central to learning and memory.

The idea that memories are stored in synaptic connections dates back to the late 19th century, when Santiago Ramón y Cajal demonstrated that the nervous system consists of individual neurons and proposed that learning might involve changes at synapses. Building on this foundation, Donald Hebb introduced a functional theory in 1949, suggesting that learning occurs through activity-dependent strengthening of synapses. This concept, known as Hebbian learning, also introduced the ideas of the “Hebb synapse” and “cell assemblies,” linking neural activity to psychological processes. The theory gained experimental support in 1966 when Terje Lømo observed long-term potentiation (LTP) in rabbit hippocampi, providing physiological evidence for synaptic strengthening. Hebbian learning continues to influence both neuroscience and artificial intelligence by inspiring artificial neural networks that emulate synaptic plasticity through adjustable connection weights.

LTP has four well-established characteristics that help explain how it contributes to memory formation:

1. **Input Specificity**: Only the synapse that receives the stimulation is strengthened, while neighboring synapses remain unchanged.
2. **Associativity**: A weak input can be strengthened if it occurs at the same time as a strong input on another synapse, allowing both to benefit from simultaneous activity.
3. **Cooperativity**: Several weak inputs that activate a neuron together can combine their effects to trigger LTP, strengthening all participating connections.

4. **Persistence**: Once established, LTP can last from minutes to months, making it a key mechanism in long-term memory storage.

Understanding these processes not only deepens our insight into brain function but also informs the development of computational models and neural network design by leveraging principles of neural plasticity and learning.

1.5.4 Learning Curve and Oblivion in Human Learning

Understanding how synaptic connections strengthen or weaken provides valuable insight into the biological basis of learning and memory. However, to fully appreciate how these mechanisms manifest in real-world behavior, it is helpful to examine empirical studies on memory retention and forgetting.

One of the most influential studies in this area was conducted by Hermann Ebbinghaus, who systematically explored how quickly newly learned information is forgotten over time. His research gave rise to the concept of the *forgetting curve*, as illustrated in Fig. 1.9.

The curve illustrates that memory retention declines rapidly if newly learned material is not reviewed. This effect is known as *oblivion*. Many individuals struggle to recall new information even a day after first encountering it. However, the graph also shows that with each repetition, the rate of forgetting decreases significantly. When the same content is reviewed after 2, 3, and 6 days, the forgetting curve becomes progressively less steep. This repeated reinforcement gradually consolidates the information into long-term memory [23].

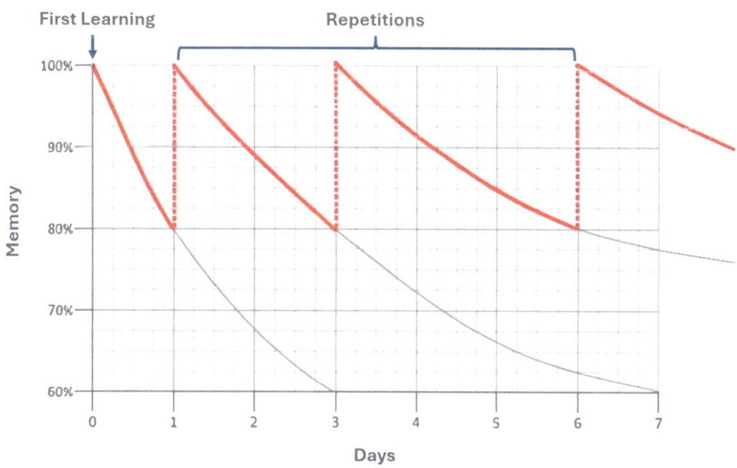

Fig. 1.9 The human learning curve according to Hermann Ebbinghaus, showing memory retention over time with and without repetition

As the saying goes, "*Repetition is the mother of learning*". The pattern is clear: when we encounter something for the first time, it is easy to forget. Upon the second exposure, it becomes harder to forget, and by the third time, forgetting becomes even more difficult. Each repetition strengthens the memory trace, increasing the likelihood of long-term retention [24].

Beyond simply preventing information loss, forgetting serves an important functional role both individually and socially [25]. It enables the brain to:

- Remove outdated knowledge and make room for new ideas.
- Discard false or irrelevant information that lacks retrieval cues.
- Prevent over-reliance on outdated strategies, thereby promoting flexibility and the motivation to acquire new knowledge.

Human learning is largely driven by repetition and the frequency of exposure to information. These behavioral patterns align with the mechanisms of synaptic plasticity discussed earlier, where repeated activation of neural pathways strengthens connections, enabling effective learning and long-term memory formation.

Insights into how learning and memory function, particularly through repetition, synaptic plasticity, and the role of forgetting, highlight the brain's remarkable adaptability. However, learning is only one aspect of a broader cognitive challenge. Another major challenge is the brain's continuous management and interpretation of the vast amount of sensory input it encounters at every moment.

To support both memory formation and real-time decision-making, the brain must go beyond retention. It must also excel at filtering. Not all information can or should be stored; much of it must be rapidly evaluated, prioritized, or dismissed. This brings us to the next level of cognitive function: how the brain processes, filters, and selects from the overwhelming stream of incoming data. The following section explores how the brain achieves this through intricate information-processing and filtering mechanisms that enable attention, pattern recognition, and cognitive efficiency in daily life.

1.6 Information Processing and Filtering

The human brain is an extraordinarily complex and sophisticated information-processing system. It continuously receives a vast influx of sensory data, such as sights, sounds, smells, tastes, and tactile inputs, which far exceed what we can consciously register. In today's information-rich environment, the brain's ability to manage and filter this constant stream is essential for avoiding cognitive overload and supporting effective decision-making.

This flow of information through the different stages of memory is explained by the Multi-Store Model of Memory. According to this model, sensory memory first receives and briefly holds incoming stimuli. When attention is applied, selected information moves into short-term memory, where it is actively maintained and

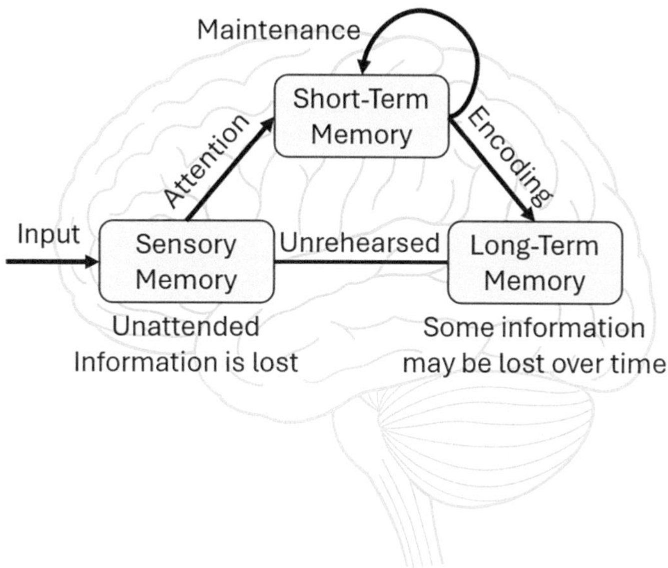

Fig. 1.10 Multi-store model of memory: the filtering of the brain

manipulated. Through processes such as rehearsal and encoding, some of this information is transferred into long-term memory for sustained storage. Each stage plays a vital role in filtering data: unattended information in sensory memory fades quickly, unrehearsed content in short-term memory is lost, and even long-term memories may weaken if not retrieved over time. These memory systems provide a structured mechanism for managing information flow, supporting human cognition, problem-solving, and survival in a constantly changing environment, as shown in Fig. 1.10.

- **Sensory Memory**: Sensory memory acts as the brain's first line of defense against information overload. It briefly stores raw input from the environment, such as visual, auditory, and other sensory data, for just a few milliseconds to a couple of seconds. This allows the brain to decide which stimuli are important enough to process further. Most sensory data is discarded at this stage, but those that capture attention are passed on to short-term memory for further evaluation.
- **Short-Term Memory**: Short-term memory temporarily holds and processes information that we are actively thinking about. It typically lasts for 15–30 s and has limited capacity, usually handling 5 to 9 items at a time. At this level, the brain filters and evaluates incoming data to determine its relevance and usefulness. Rehearsal and focused attention can help maintain information in short-term memory or transfer it to long-term memory.
- **Long-Term Memory**: Long-term memory stores information over extended periods, from hours to a lifetime. It includes both conscious memories (e.g., facts and events) and unconscious skills (e.g., riding a bicycle). This stage involves the consolidation of meaningful and frequently accessed information. Filtering at this

level ensures that only valuable and reinforced knowledge is retained, while irrelevant or outdated information is forgotten. This dynamic process is essential for adaptive learning, decision-making, and knowledge retention.

By coordinating sensory, short-term, and long-term memory systems, the brain filters vast amounts of data, recognizes patterns, integrates experience, and enables us to focus, learn, and act efficiently in a complex world.

1.6.1 Sensory Memory

Human perception relies on five senses: vision, hearing, touch, taste, and smell. Each sense utilizes specialized neural cells, called receptors, to detect specific environmental stimuli and transmit signals to the brain. Despite processing different types of stimuli, all sensory systems follow a common principle: early sensory filtering. This process begins in peripheral structures, such as the retina, cochlea, or skin receptors, where initial processing occurs. Signals then travel through subcortical regions before reaching the neocortex for higher-order analysis.

The neocortex has a uniform, layered structure throughout the brain. This architectural consistency enables the brain to organize incoming sensory information into meaningful spatial and temporal patterns, which are essential for perception, recognition, and prediction.

- **Spatial patterns**: Signals received from photoreceptors in the retina (for vision).
- **Temporal patterns**: Changes over time, such as auditory patterns or saccadic eye movements.

For example, in the visual system, the eyes perform unconscious saccadic movements approximately three times per second, shifting the gaze from one point to another. Each saccade causes abrupt changes in retinal input, meaning the visual information received by the brain is constantly being updated. Despite these frequent shifts, we perceive a continuous and stable visual world rather than a chaotic blur. This perceptual stability is made possible by the neocortex's ability to process spatial and temporal patterns effectively. Figure 1.11 illustrates this process: the image shows jumping of the eyes (saccades) when perceiving (or rather scanning) a face (or other image), demonstrating how the brain integrates multiple viewpoints into a coherent whole.

Since the brain receives vast amounts of raw sensory data, only a fraction reaches conscious awareness. Several key structures facilitate this sensory filtering process:

Thalamus: Acts as a "sensory gatekeeper," filtering incoming signals before they reach the cortex.

Primary Sensory Cortices (V1, A1, S1): Detect fundamental features such as edges, sounds, and textures while discarding redundant information.

Reticular Activating System (RAS): Regulates attention by prioritizing significant stimuli, such as recognizing one's name in a noisy environment.

Fig. 1.11 Saccades during scanning a face through the eyes (saccades)

This filtering mechanism ensures that only relevant sensory information reaches higher cognitive levels, enhancing perception and awareness. The specific sensory processes for each sensory modality are as follows.

1.6.1.1 Vision

The visual system is the most complex sensory system, processing visual information in parallel through the retina, thalamus, and the occipital lobe of the neocortex. Vision begins when retinal photoreceptors capture photons, converting them into electrical signals. The retina and brain perform complex analyses of this information, with retinal neurons extracting essential features like edges, motion, and color contrasts. There are approximately 15 visual pathways that unconsciously process visual input for functions such as controlling pupil diameter and circadian rhythms.

Photoreceptors do not send the entire image directly to the brain. Instead, they interact with other retinal neurons to extract specific features before transmitting the processed signals. The retina contains over 100 million neurons, but only about one million axons relay visual information to the brain, requiring an efficient filtering process to prioritize essential details while discarding redundant background noise.

At the subcortical level, the retina sends its output to the dorsal lateral geniculate nucleus (dLGN) of the thalamus, which acts as a sensory gatekeeper. The dLGN filters and modulates responses based on attention and relevance, ensuring that only the most salient visual data reach higher-order processing centers. From the thalamus, signals travel to the primary visual cortex (V1) in the occipital lobe, where neurons specialize in detecting motion, orientation, and spatial features. These cells fire action potentials only when they detect a line or edge of a particular orientation or movement in the visual field.

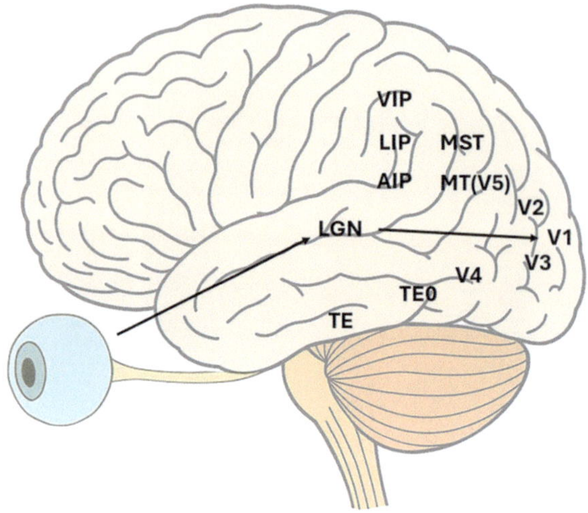

Fig. 1.12 Pathways in the visual system

The visual system contains many areas that process hierarchically, as illustrated in Fig. 1.12. After V1, the visual system provides two major pathways: the dorsal and ventral streams. The dorsal stream processes from the V2/V3 include MT and MST in the parietal lobe that responds to image movement. The dorsal stream refers to *where* for visually guided behavior such as catching or kicking a ball or running through obstacles. The ventral stream comprises V4, TE, and TEO in the inferior temporal lobe. The ventral pathway is often referred to *what* as it responds to particular patterns or colors such as faces.

1.6.1.2 Hearing

The auditory system processes sound through air vibrations. The outer ear captures sound waves and directs them to the middle ear, where vibrations in the ossicles (middle ear bones) convert mechanical energy into fluid waves within the cochlea. Inside the cochlea, hair cells function as mechanoreceptors, encoding frequency through spatial positioning along the basilar membrane and amplitude through variations in action potentials.

To prevent sensory overload and enhance auditory perception, early sensory filtering occurs at multiple levels:

The cochlea conducts initial frequency filtering through tonotopic organization, ensuring that specific sound frequencies activate designated regions along the basilar membrane. Signals are then transmitted to the brainstem nuclei, where background noise is suppressed and key auditory patterns, such as speech and sudden environmental sounds, are emphasized. The medial geniculate nucleus (MGN) of the thalamus functions as a sensory gatekeeper, selectively refining auditory input based on attention and relevance. Finally, auditory signals reach the primary auditory cortex (A1) in the temporal lobe, where neurons specialize in processing pitch, rhythm, and spatial location.

1.6.1.3 Touch

The sense of touch relies on specialized mechanoreceptors in the skin that detect pressure, vibration, and stretching, while free nerve endings respond to temperature and pain. These sensory signals are transmitted through the spinal cord, where initial processing occurs to filter redundant or less critical information before reaching higher brain centers.

To enhance perceptual efficiency, early sensory filtering takes place at multiple stages. In the spinal cord, some sensory inputs are modulated through reflex pathways or inhibitory interneurons, preventing excessive or unnecessary stimuli from overwhelming the system. The processed signals then ascend to the ventral posterior nucleus of the thalamus, which serves as a sensory gatekeeper, selectively refining tactile information based on its relevance.

From the thalamus, the filtered signals are relayed to the somatosensory cortex in the parietal lobe, where neurons specialize in decoding spatial and temporal patterns of touch. This hierarchical processing allows the brain to efficiently interpret and respond to tactile sensations, ensuring precise perception of texture, pressure, and pain while maintaining sensory balance.

1.6.1.4 Smell

The sense of smell is unique among sensory systems because it bypasses the thalamus, allowing odor signals to reach higher brain regions more directly. Olfactory receptors in the nasal cavity detect airborne odor molecules, converting them into neural signals. Humans have approximately 10 million olfactory receptors, enabling the detection of a vast range of odors.

Unlike other senses, early sensory filtering in the olfactory system occurs primarily in the olfactory bulb, where redundant signals are reduced, and significant odor patterns are textitasized before being transmitted further. The processed signals are then projected directly to the neocortex for higher-order processing and to the amygdala, a key region involved in emotion and memory. This direct connection between olfaction and the limbic system explains why certain smells can evoke strong emotional responses or trigger vivid memories.

1.6.1.5 Taste

Taste perception is mediated by taste receptors located in the papillae on the tongue, as illustrated in Fig. 1.13. These receptors detect five primary taste modalities: sweet,

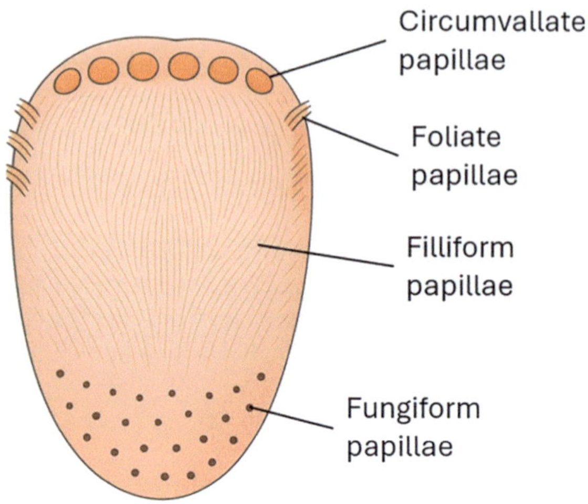

Fig. 1.13 Receptors in the tongue

sour, salty, bitter, and umami. Chemical stimuli from food are converted into neural signals that travel to the nucleus of the solitary tract (NST) in the brainstem, initiating early sensory processing.

At this stage, redundant or less relevant taste information is modulated before being relayed to the ventroposteromedial thalamus (VPM), which acts as a sensory gatekeeper, refining the signals based on intensity and relevance. The filtered signals are then transmitted to the gustatory cortex, where further processing allows for taste discrimination, integration with olfactory and texture cues, and the generation of conscious taste perception. This hierarchical filtering system ensures that the most salient taste information contributes to food evaluation, appetite regulation, and flavor perception.

1.6.2 Memory-Based Filtering

Humans are naturally skilled at solving problems and planning for survival, but this process requires a conscious state to maintain context and retain relevant information long enough to complete the task at hand. Short-term memory plays a crucial role in this, as it holds information temporarily before the brain shifts attention to new challenges. To ensure that important information endures beyond the present moment and supports future decision-making, it must be transferred to long-term memory. This more durable storage system enables the brain to selectively retain knowledge that is meaningful or frequently encountered, allowing for sustained retention and efficient filtering and prioritization of information over time.

1.6.2.1 Short-Term Memory

Short-term memory (STM) functions as the brain's temporary workspace, essential for retaining and manipulating information required for immediate cognitive tasks and problem-solving. It enables individuals to hold relevant details long enough to carry out actions such as mental calculations, short conversations, or recalling directions. This system is constrained by its limited capacity, typically 7±2 items, and its fragile duration, with unreinforced information often fading within 15 to 30 s.

Short-term memory is primarily supported by the prefrontal cortex, especially the dorsolateral region, which keeps important information active in the brain through continuous electrical signaling known as persistent neuronal firing. This allows us to temporarily hold information even after the original input is gone. Brainwave patterns, particularly the synchronization of theta (slow) and gamma (fast) waves, help regulate how many items we can keep in mind at once. Research shows that longer theta cycles can improve memory accuracy by allowing more time to process each piece of information.

At the molecular level, cyclic AMP (cAMP) plays a crucial role by enhancing the release of neurotransmitters, which strengthens communication between brain cells. These temporarily boosted connections, called synapses, help support the short-term retention of information. Although these changes do not last long, they are essential for holding information just long enough to complete tasks like doing math in your head or remembering a name during a conversation.

The hippocampus also contributes significantly to short-term memory, particularly in associating sensory inputs with context. For example, seeing a red bird triggers activation in a specific hippocampal region, helping the brain encode that experience into a coherent neural pattern (Fig. 1.14). Beyond encoding, the hippocampus acts as a communication bridge by projecting activity back to the neocortex to recreate and strengthen original neural connections. This is an important step in transitioning short-term memories into long-term storage through rehearsal and repetition.

Short-term memory plays another important role as a cognitive filter. Because it has limited capacity, the brain must constantly decide which pieces of information to keep and which to ignore. This selective process helps prevent mental overload and allows us to stay focused on what is most important at any given moment.

In everyday situations, short-term memory supports many essential tasks. For example, mental arithmetic requires holding numbers and intermediate steps in mind, such as when calculating 23 times 4. Navigating to a destination involves remembering a sequence of turns, and social interactions often rely on briefly remembering a person's name right after being introduced. However, short-term memory is also fragile. A brief distraction, such as a sudden noise or interruption, can cause this information to disappear, showing how easily it can be lost.

Experimental studies have provided valuable insights into the biological basis of short-term memory. A commonly used method involves T-maze tasks, in which rats or mice are trained to remember which arm of a T-shaped maze they previously entered in order to make the correct choice in the next trial. These tasks have highlighted the importance of specific brain regions, particularly the hippocampus and

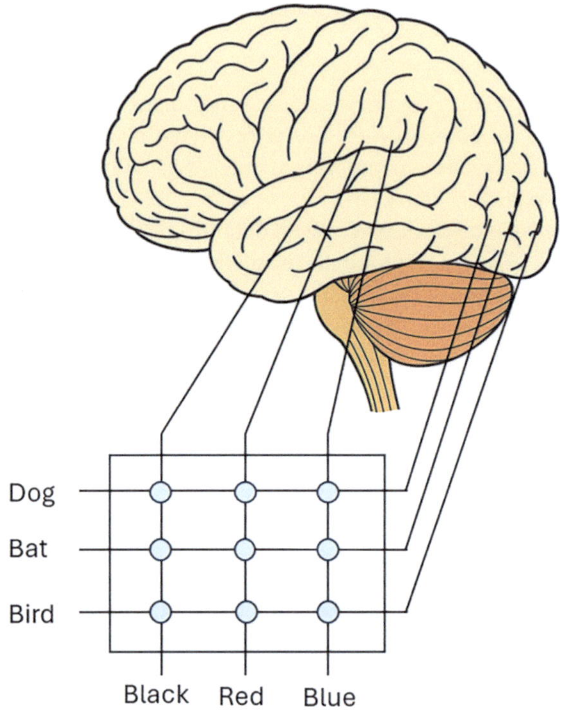

Fig. 1.14 Simplified hippocampus structure

the striatum. When these areas are intact, the animals are able to recall their previous actions even after a short delay. However, when these regions are damaged, especially during delayed alternation tasks that include a pause between trials, the animals often fail to remember their last choice and show impaired performance. These findings emphasize the essential role of the hippocampus and striatum in supporting short-term memory.

These results show that short-term memory is not just a passive system for holding information. Instead, it actively helps the brain connect recent sensory experiences with longer-term learning, making it a key part of decision-making and future behavior. However, STM is also easily disrupted, especially when multiple tasks or distractions compete for attention. This makes it clear that healthy brain circuits are crucial for STM to function effectively and that even small disruptions can lead to noticeable memory problems.

1.6.2.2 Long-Term Memory

Long-term memory (LTM) serves as the brain's enduring storage system, capable of retaining information for years or even a lifetime. Unlike the transient nature of short-term memory, LTM is built through repeated exposure, emotional salience, or strong contextual associations. It operates through complex biological mechanisms involving both structural changes in the brain and dynamic processing during periods of rest, particularly sleep.

One of the most critical processes in forming long-term memory is memory consolidation, which transforms short-term experiences into stable, retrievable memory traces. The hippocampus plays a central role in this process, acting as a sorting hub that temporarily stores new information and, during REM and slow-wave sleep, replays these experiences. These reactivations, such as a rat mentally retracing a maze, strengthen neural pathways through long-term potentiation (LTP), a mechanism in which repeated activation of synapses enhances their strength. Over time, these memories are distributed to the neocortex, where they settle into specialized regions based on content. For instance, visual memories embed in the occipital lobe, while auditory ones store in the temporal lobe, forming a decentralized yet interconnected memory network.

Biochemically, consolidation is driven by protein synthesis, including enzymes like PKMζ, which stabilize the structural changes that allow for long-term retention. Additionally, theta-gamma oscillations, which help regulate short-term memory, continue to support the reactivation and encoding of long-term memories during sleep. Emerging studies have shown that cyclic-AMP pathways enhance neurotransmitter release and support the molecular changes necessary for memory stabilization.

Long-term memory (LTM) is generally divided into two main subsystems: explicit (declarative) and implicit (non-declarative) memory. Explicit memory involves conscious recall of information. It includes episodic memory, which stores personal experiences such as attending a graduation ceremony, and semantic memory, which contains general knowledge like the names of capital cities. The hippocampus plays a central role in this system, especially in forming and retrieving new episodic and semantic memories.

In contrast, implicit memory operates without conscious awareness. It includes skills and habits, such as riding a bicycle, as well as emotional responses like fear, which are often regulated by the amygdala. These memory systems can function independently. For example, individuals with damage to the hippocampus may lose the ability to recall personal events but still retain motor skills. This separation highlights the specialized and distributed nature of long-term memory across different brain regions.

Long-term memory is not only about retention; it is also about filtering and efficiency. During sleep, the brain strengthens meaningful connections while pruning away less useful ones, preserving only essential information for future use. This selective process enhances cognitive efficiency and ensures that knowledge critical for survival or task performance is retained. Mechanisms such as emotional salience, driven by dopamine release during significant or rewarding experiences, further enhance the encoding of memories that are personally relevant.

The process of forgetting, though often seen as a limitation, is essential for long-term memory management. It prevents cognitive overload, enables faster access to relevant information, and supports adaptability. For a discussion of how forgetting functions biologically and behaviorally, including the foundational work of Hermann Ebbinghaus and the forgetting curve, please refer to Sect. 1.5.3, Learning Curve and Oblivion in Human Learning.

Recent research is reshaping our understanding of long-term memory. It was once believed that all long-term memories must first pass through short-term memory. However, a 2024 study [26] found that some long-term memories can form independently, through CaMKII-independent pathways. These pathways bypass the enzyme CaMKII, which was previously thought to be essential for memory stabilization. This reveals that the brain may use alternative mechanisms, especially during stress, trauma, or rapid learning.

Another important discovery is the concept of "silent engrams" [27], which are memory traces stored in the brain that remain inactive and inaccessible under normal conditions. These dormant memories can be reactivated later through specific triggers, such as artificial stimulation or strong emotional cues. This suggests that memories we think are forgotten may not be lost, but simply hidden from conscious access. These findings highlight the brain's adaptability and offer new insights that could inspire more flexible and resilient memory architectures in artificial intelligence systems.

The layered architecture of memory, where short-term memory provides a transient workspace and long-term memory acts as a durable archive, empowers humans to adapt, learn, and thrive in complex environments. This dynamic system ensures a balance between immediate demands and lasting knowledge, allowing us to draw upon past experiences while remaining open to future learning.

1.7 Conclusion

Understanding the human brain is a crucial first step in bridging the gap between biological and artificial intelligence. This chapter began by comparing the brain's capabilities with those of current artificial systems. Section 1.1 established the brain's superior energy efficiency, parallel processing, and fault tolerance, highlighting why it remains the gold standard for intelligence. Section 1.2 then traced the brain's evolutionary development from simple nerve nets to complex mammalian systems, emphasizing its refined architecture and hierarchical structuring.

Next, Sect. 1.3 detailed the functional organization of brain components like the cerebellum, hippocampus, thalamus, and limbic system, showing their integrated roles in processing sensory inputs, regulating emotion, and controlling behavior. This led to Sect. 1.4, which focused on the neocortex, the uniquely mammalian structure responsible for higher-order cognition, reasoning, and decision-making. It integrates structure and function to drive goal-directed behavior.

Section 1.5 shifted to the underlying neural processes, examining neuron structure, action potentials, and synaptic plasticity to explain how information is transmitted, learned, and retained. It also touched on memory dynamics and the forgetting curve.

Finally, Sect. 1.6 explained the brain's hierarchical memory system, detailing how sensory, short-term, and long-term memory filter and prioritize vast amounts of sensory input based on relevance, essential for cognitive efficiency and directly relevant to AI challenges in attention and memory.

This chapter set the stage by linking brain structure, function, and learning mechanisms. With this foundation in place, we now turn to the neocortex, the most advanced region of the brain associated with intelligence. The next chapter explores its structure and defining characteristics and examines how these biological principles have inspired computational models. Particular attention is given to the ideas of Jeff Hawkins, whose framework is based on key attributes of the neocortex, including hierarchical organization, sequence learning, abstraction, and generalization. These concepts underpin more brain-like approaches to artificial intelligence.

References

1. Thompson, R.F.: The Brain: A Neuroscience Primer. Worth Publishers, New York (2000). 9780716732266
2. Intel Corporation: Intel Builds World's Largest Neuromorphic System to Enable More Sustainable AI. (2025). https://newsroom.intel.com/artificial-intelligence/intel-builds-worlds-largest-neuromorphic-system-to-enable-more-sustainable-ai. Accessed 4 May 2025
3. Akopyan, F., et al.: TrueNorth: design and tool flow of a 65 mW 1 Million Neuron programmable neurosynaptic chip. IEEE Trans. Comput. Aided Des. Integr. Circuits Syst. **34**(10), 1537–1557 (2015). https://doi.org/10.1109/TCAD.2015.2474396
4. Agoston, D.V.: Of artificial intelligence, machine learning, and the human brain. Celebrating Miklos Palkovits' 90th birthday. Front. Neuroanatomy **18**, 1374864 (2024). https://doi.org/10.3389/fnana.2024.1374864. PMID: 38764486; PMCID: PMC11099251
5. Plank, P., et al.: A Long Short-Term Memory for AI Applications in Spike-based Neuromorphic Hardware (2021). arXiv:2107.03992. https://arxiv.org/abs/2107.03992
6. IBM leverages SoC tech in supercomputer design. (2005). https://www.design-reuse.com/news/202505350-ibm-leverages-soc-tech-in-supercomputer-design/. Accessed 4 May 2025
7. Vangal, S.R., Howard, J., Dighe, S., et al.: An 80-Tile 1.28TFLOPS Network-on-Chip in 65nm CMOS. In: IEEE International Solid-State Circuits Conference Digest of Technical Papers (ISSCC), pp. 98–589 (2007). https://doi.org/10.1109/ISSCC.2007.373159
8. Levy, W.B., Calvert, V.G.: Communication consumes 35 times more energy than computation in the human cortex, but both costs are needed to predict synapse number. Proc. Natl. Acad. Sci. USA **118**(23), e2008173118 (2021). https://doi.org/10.1073/pnas.2008173118
9. Arute, F., et al.: Quantum supremacy using a programmable superconducting processor. Nature **574**, 505–510 (2019). https://doi.org/10.1038/s41586-019-1666-5
10. Neyigapula, B.S.: Quantum neural networks: paving the way for next-generation machine learning. Int. J. Artif. Intell. Mach. Learn. **4**(2), 92–105 (2024). https://doi.org/10.51483/IJAIML.4.2.2024.92-105
11. Rebentrost, P., Mohseni, M., Lloyd, S.: Quantum support vector machine for big data classification. Phys. Rev. Lett. **113**(13), 130503 (2014). https://doi.org/10.1103/PhysRevLett.113.130503
12. Google Quantum AI: TensorFlow Quantum: Hybrid Quantum-Classical Machine Learning. (2020). https://www.tensorflow.org/quantum. Accessed 4 May 2025
13. Microsoft Corporation: Microsoft's Majorana-1 chip carves new path for quantum computing. (2024). https://news.microsoft.com/source/features/innovation/microsofts-majorana-1-chip-carves-new-path-for-quantum-computing/. Accessed 4 May 2025

14. Pasqal Team: Quantum Computing: Rethinking Energy Consumption (2025). https://www.pasqal.com/blog/quantum-computing-rethinking-energy-consumption/. Accessed 4 May 2025
15. Martin, M.J., Hughes, C., Moreno, G., et al.: Energy use in quantum data centers: scaling the impact of computer architecture, qubit performance, size, and thermal parameters. Nat. Renew. Energy Labor. Tech. Rep. NREL/TP-2C40-77200 (2021). https://doi.org/10.2172/1880806
16. Simmons, P.J., Young, D.: Nerve Cells and Animal Behaviour. Cambridge University Press (2010). ISBN: 9780511782138
17. Dent, S.: Researchers take full control of cockroach's movement, turn it into a wireless sensor. (2012). https://www.engadget.com/2012-09-06-researchers-guide-cockroach-wirelessly.html. Accessed 4 May 2025
18. Erickson, J.C., Herrera, M., Bustamante, M., Shingiro, A., Bowen, T.: Effective stimulus parameters for directed locomotion in madagascar hissing cockroach biobot. PLoS ONE **10**(8), e0134348 (2015). https://doi.org/10.1371/journal.pone.0134348
19. Martin, J.P., Guo, P., Mu, L., et al.: Central-complex control of movement in the freely walking cockroach. Curr. Biol. **25**(21), 2795–2803 (2015). https://doi.org/10.1016/j.cub.2015.09.044
20. Yang, Z.: The principle of cortical development and evolution. Neurosci. Bull. **41**, 461–485 (2025). https://doi.org/10.1007/s12264-024-01259-2
21. Majoral, D., Zemmar, A., Vicente, R.: A model for time interval learning in the Purkinje cell. PLoS Comput. Biol. **16**(2), e1007601 (2020). https://doi.org/10.1371/journal.pcbi.1007601
22. Hebb, D.O.: The Organization of Behavior: A Neuropsychological Theory. Wiley, New York (1949)
23. Ebbinghaus, H.: Memory: A Contribution to Experimental Psychology. Teachers College, Columbia University: Translated by Henry A, p. 1913. Ruger and Clara E, Bussenius (1885)
24. Yang, J., et al.: Effects of repetition learning on associative recognition over time. Front. Hum. Neurosci. **12**, 274 (2018). https://doi.org/10.3389/fnhum.2018.00274
25. Assmann, A.: Social forgetting: a systems-theory approach. In: Erll, A., Nünning, A. (eds.), Cultural Memory Studies: An International and Interdisciplinary Handbook, pp. 181–190. De Gruyter (2008)
26. Shin, M.E., Parra-Bueno, P., Yasuda, R.: Formation of long-term memory without short-term memory revealed by CaMKII inhibition. Nat. Neurosci. **28**, 35–39 (2024). https://doi.org/10.1038/s41593-024-01831-z
27. Roy, D.S., Arons, A., Mitchell, T.I., Pignatelli, M., Ryan, T.J., Tonegawa, S.: Silent memory engrams as the basis for retrograde amnesia. Proc. Natl. Acad. Sci. **114**(46), E9972–E9979 (2017). https://doi.org/10.1073/pnas.1714248114

Chapter 2
The Neocortex

Building upon the foundation laid in Chap. 1, which explored the brain's evolution, structure, and cognitive efficiency, in this chapter focuses specifically on the neocortex, the brain region most associated with higher-order cognition and intelligence. This chapter is essential for understanding how the neocortex's consistent architecture and dynamic functionality provide a blueprint for artificial intelligence. While Chap. 1 highlighted the brain's remarkable efficiency and adaptability, in this chapter delves deeper into the neocortex's ability to process complex sensory input, form abstract representations, and guide intelligent behavior. These biological insights set the stage for modeling intelligent systems based on the brain's most powerful learning mechanisms.

The chapter first introduces four fundamental properties of the neocortex: structural uniformity, invariant representation, hierarchical organization, and auto-associative memory (Sect. 2.1). These features enable robust pattern recognition and contextual prediction. It then presents Jeff Hawkins' theories (Sect. 2.2), emphasizing the brain's use of shared learning algorithms across all functions, the role of temporal memory and sequence prediction, and the integration of sensory and motor signals. The Thousand Brains Theory proposes that perception arises from consensus among thousands of mini-models (cortical columns), each building partial representations of reality in parallel.

In the latter part of the chapter, attention shifts to how these neocortical mechanisms translate into machine intelligence (Sect. 2.4). This section introduces Hierarchical Temporal Memory (HTM) as a brain-inspired computational model that captures the neocortex's core principles, such as hierarchy, temporal learning, prediction, and adaptability. It explains how HTM employs sparse representations, continuous online learning, and structural parallels to cortical columns to process information in a way that closely mirrors biological intelligence. A brief conclusion can be found in Sect. 2.5.

T. Soisoonthorn and H. Unger, *A Brain-Inspired Approach to Natural Language Processing*, Studies in Computational Intelligence 1225,
https://doi.org/10.1007/978-3-032-00014-9_2

2.1 Anatomical Structure

The neocortex is a fundamental component of the mammalian brain, responsible for higher cognitive functions such as perception, language, abstract reasoning, and decision-making. Despite handling a wide range of tasks, it operates based on a remarkably consistent set of principles and structures. This uniformity enables it to recognize patterns, learn from experience, and adapt to new situations, making it a highly efficient system for processing information.

The following sections explore four fundamental attributes that define the structure and function of the neocortex. These attributes reveal how it organizes and processes information, supporting complex thought, learning, and adaptation. A deeper understanding of these mechanisms provides valuable perspectives on both biological intelligence and the advancement of artificial systems.

2.1.1 Uniformity of the Neocortex

A noteworthy feature of the neocortex is the uniformity of its fundamental structure. It is composed of a single repeating mechanism, with the cortical column serving as the basic functional unit. A cortical column is a vertically organized group of neurons that spans all six layers of the neocortex and is responsible for processing a specific type of input. Each column operates like a miniature processor, analyzing a small segment of sensory data independently. Estimates suggest that each column contains approximately 100 neurons, resulting in roughly 100 million columns in the human brain. Each column represents a distinct entity yet functions in coordination with the others.

Although different regions of the neocortex are specialized to recognize and process specific patterns depending on their location, the overall architecture remains consistent. If the normal processing pathway is disrupted, such as by injury, other regions can often compensate. This adaptability, known as plasticity, is well documented in cases where patients with brain damage from strokes or trauma relearn lost skills by recruiting unaffected areas of the neocortex. Such findings provide strong support for the view that the neocortex is fundamentally uniform, with each column capable of performing similar functions regardless of its location.

2.1.2 Invariant Representation

Invariant representation in the brain means that each neural unit represents one specific thing, even if the thing is perceived in different ways. For example, when we see a friend's face in varying lighting conditions, angles, or while they are walking, the brain still recognizes them as the same person. In an experiment that monitored the activity of neurons in different layers of the brain, it was found that while the pattern of activity in V1 changed constantly as patterns were received, the neurons in the face recognition area of the neocortex remained stable.

2.1.3 Hierarchy

Each area of the thalamus system has a large projection up to the neocortex. The neocortex also projects to secondary, tertiary, and higher areas within itself. Each of these areas also projects back to the area that is projected to it, forming a complex loop of information exchange. Neurons in higher layers along these information flows typically respond to more abstract or selective stimuli than the neurons in the lower areas. For example, when reading the phrase "The Human Brain...," the visual system sends signals from the retina to the thalamus, which relays the signals to the visual cortex area, V1. V1 lowers stroke detectors and sends the signals to other areas in the neocortex to activate in higher levels until the word detectors in Wernicke's area recognize the phrase. This recognition occurs through a hierarchy of information processing, with lower layers handling lower features and higher layers working with more abstract concepts. Understanding the innate hierarchical nature of reality is a unique ability of mammals and humans. Animals without a neocortex are incapable of understanding hierarchies (Fig. 2.1).

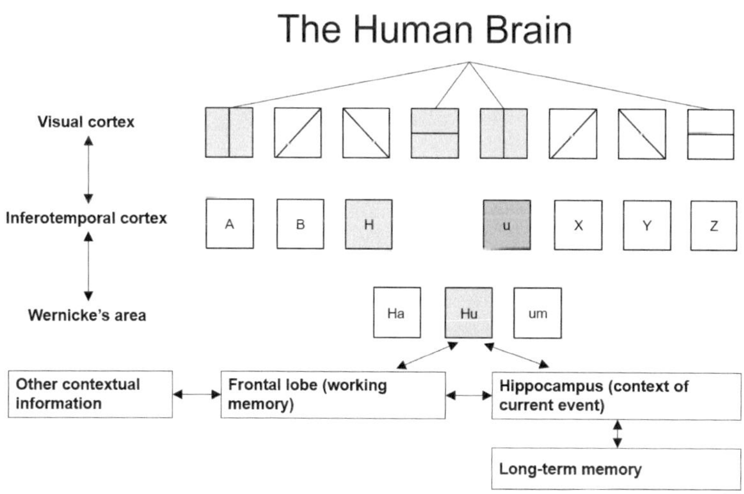

Fig. 2.1 Think in hierarchical nature of the neocortex

2.1.4 *Auto-Associated Memory*

Auto-associated memory is the brain's ability to recognize a whole pattern from just a part of it. This mechanism combines elements of both episodic memory and semantic memory. As information moves from lower to higher levels in the brain, the connections between them are weighted based on how important each part of the pattern is. The more important a part is, the stronger its connection. This means that even if some parts of the pattern are missing or unclear, the brain can still recognize the whole based on the key elements that remain.

An important aspect to consider is that when neurons project signals to other neurons or brain areas, they often also send feedback signals. This backward projection supports the brain's ongoing effort to predict what comes next. For example, when reading the sentence "The cat is eatin," our brain automatically anticipates the next letter to be "g." Similarly, when we take a shower, we think at a higher, abstract level about the act of showering, while the detailed actions unfold automatically and unconsciously because they have already been learned at lower levels. This feedback mechanism enables prediction. However, when events do not align with expectations, the anterior cingulate cortex becomes active. It monitors discrepancies between planned actions and actual outcomes, especially during complex tasks or errors that require conscious attention, adaptation, and learning.

The thalamus also plays a key role in this feedback system. As a central relay station, it not only transmits sensory input to the cortex but also helps regulate the flow of feedback between cortical regions. It participates in both bottom-up and top-down communication, enabling higher-level areas to influence how new input is processed. This coordination is especially critical in sequential tasks such as language comprehension, movement, or sensory perception, where context and timing determine what should happen next. By linking previous inputs to ongoing activity, the thalamus supports the continuity of patterns and assists in reconstructing incomplete or noisy information. Finally, because neurons are inherently noisy and input may be partially degraded, higher cortical layers can use feedback from similar lower-level patterns to reconstruct missing or incorrect elements. This ability to "fill in the blanks" helps the brain maintain stable perception and memory, even under uncertain conditions.

2.2 The Fundamental Ideas of Hawkins

Jeff Hawkins, an American inventor, neuroscientist, and entrepreneur, has significantly contributed to our understanding of intelligence and brain function. To fully grasp his groundbreaking theories on intelligence and the neocortex, it is essential to first understand the context and historical importance of his work.

In the early 2000s, while AI research focused on specialized algorithms, Hawkins took a different path. Drawing on his tech background, he founded Numenta in 2005 to develop machine intelligence inspired by the neocortex.

Hawkins proposed in his 2004 book On Intelligence [1] that building intelligent machines might require a deeper understanding of how the human brain works, particularly the neocortex. This perspective contrasts with traditional AI approaches, which aim to achieve intelligent behavior without necessarily replicating biological processes.

2.2.1 *Common Algorithms*

A key insight from neuroscience builds on Vernon Mountcastle's 1978 [2] discovery that the neocortex operates using the same computational principles across all of its functions. Whether processing visual, auditory, tactile, or linguistic information, the underlying neural algorithms remain consistent. This structural and functional uniformity suggests that the neocortex implements a universal learning algorithm that can be applied to any type of sensory input or behavioral output.

This finding has significant implications for artificial intelligence. Rather than designing separate algorithms for individual cognitive tasks such as vision, speech, or motor control, researchers are increasingly focused on identifying and replicating the common principles used by the neocortex. This strategy may support the development of more general and adaptable machine intelligence that is capable of learning and performing a wide range of tasks using the same foundational mechanisms.

Support for this idea is reinforced by neuroscience research, particularly studies on how the brain adapts following injury. When an individual experiences a traumatic brain injury (TBI) or a stroke, the brain does not simply stop functioning in the damaged areas. Instead, it actively reorganizes itself to compensate for the loss. For example, cognitive training after TBI has been shown to reduce the brain's tendency to isolate functions within specific regions, a property referred to as modularity. As a result, the brain becomes more integrated, allowing different regions to collaborate more effectively. Functional MRI studies have shown that undamaged areas increase communication with other parts of the brain. This is reflected in higher participation coefficients and improved overall network efficiency. Healthy regions are often able to take over some of the functions previously managed by the damaged areas.

A similar process occurs during stroke recovery. Brain regions surrounding the damaged site, known as peri-infarct regions, frequently rewire themselves to perform sensory or motor functions that were disrupted. These examples demonstrate how the neocortex, composed of repeating structures governed by shared learning rules, allows one area to assume the role of another. This flexibility illustrates the brain's capacity to generalize and reuse common algorithms across different functions, offering valuable guidance for the development of brain-inspired artificial intelligence.

2.2.2 *Hierarchical Organization and Memory*

Human intelligence depends not just on raw brain power, but on how well the brain can store experiences and use them to make predictions. One of the most important reasons it can do this so effectively is the structure of the neocortex, which makes up about 80% of the human brain. This large area is organized in a hierarchy, meaning it processes information in layers, starting from simple details and building up to complex understanding.

At the lower levels, the brain handles raw sensory data, like light, sound, or touch. As this information moves up through the layers, the brain begins to recognize patterns, understand relationships, and store the information as structured knowledge. The higher levels deal with more abstract ideas, like recognizing a person's face, understanding the meaning of words, or guessing someone's intention. This layered structure helps the brain not only recognize and classify information but also understand it in context and predict what might happen next.

This hierarchical design brings many advantages. First, it allows the brain to break down complex problems into simpler parts, making it easier to understand and react to new situations. Second, it helps with abstraction and generalization. For example, once the brain learns what a "dog" is, it can recognize all kinds of dogs without needing to relearn each one. Third, it makes the brain more robust to noise or missing information; even if part of what we see or hear is unclear, the brain can often fill in the gaps. Fourth, it supports prediction and planning, helping us anticipate what comes next based on experience. Finally, it allows for reusability. Patterns learned in one situation, such as shapes or sounds, can be applied to many others, making learning faster and more flexible.

Another key benefit is efficient memory storage. The brain avoids repeating the same information by reusing lower-level features across different experiences. It stores general building blocks and combines them in different ways, instead of keeping every memory separately. Abstract concepts at higher levels act like compressed versions of many specific events. Also, because the brain is constantly predicting what will happen next, it can ignore familiar information and focus only on what's new or surprising. This helps save both space and energy, allowing the brain to store a huge amount of useful knowledge in an efficient way.

2.2.3 *Temporal*

Intelligence depends heavily on how we understand time. The brain does not see the world as separate moments but as a continuous flow. It learns by recognizing how events are connected over time, which helps us remember sequences, understand order, and predict what might happen next.

The neocortex is especially good at learning patterns that occur in a specific order across different timescales. For example, when you hear the beginning of a familiar song, you can often guess what comes next. This happens because your brain has learned the sequence through repeated experience. This ability allows us to understand speech, move smoothly, and anticipate actions in daily life.

Our sense of time is spread throughout the neocortex. Its repeating structure, made of columns, stores both single patterns and sequences. These columns are arranged in layers. Lower layers detect quick and simple signals like sounds or edges of shapes. Higher layers capture slower and more complex patterns like complete sentences or the meaning of a conversation. This structure enables the brain to move beyond reacting to the present moment. By recognizing and storing patterns across time, it supports reasoning, future planning, and decisions based on anticipated outcomes.

2.2.4 Prediction

From these time-based memories, the brain constantly forms predictions. It's always trying to guess what will happen next, using past experiences as a guide. This process, anticipating the future based on the past, is considered a core aspect of intelligence. The brain doesn't just react to what it sees or hears; it actively forecasts upcoming events and adjusts its understanding based on how things actually play out.

This predictive ability is present in all areas of perception and thought. When you listen to music, your brain is already anticipating the next note before it plays. When you're watching a moving object, your visual system is predicting where it will go. As you speak or read, your language system is guessing the next word or phrase. These predictions allow us to respond quickly, stay oriented in fast-changing situations, and fill in missing information when things are unclear.

A well-known example of prediction in action can be seen in modern language models such as GPT. These systems generate coherent and contextually relevant text by predicting the next word in a sequence based on the words that came before. This type of next-step prediction, grounded in learned patterns and context, demonstrates how anticipating future elements in a sequence enables fluent communication and adaptive behavior.

In the brain, prediction works through learned sequences stored in different regions. Special pathways, such as thalamic circuits, help maintain the order of events, allowing the brain to play back what it expects to happen next. Lower brain areas focus on short-term, fast-changing details, while higher areas form predictions over longer time spans and more abstract ideas.

There is also a constant exchange between sensory input and expectation. Information flows upward from the senses and downward from predictions. When something unexpected happens, the brain updates its model. When predictions are correct, they reinforce what the brain has learned. This feedback loop, of predicting, sensing, adjusting, and refining, happens all the time and is how we learn from experience and build a more accurate understanding of the world.

2.2.5 *Sensory-Motor Integration*

The neocortex plays a key role in connecting what we sense with how we move. The brain does not just receive sensory input passively. Instead, it actively interprets that input by considering how the body is moving at the same time. For example, when your eyes move, the visual scene changes. But you still perceive the world as stable because your brain takes your eye movements into account.

This close link between sensing and movement is essential for accurate perception. To fully understand an object or environment, we need to interact with it. We move our eyes to focus on details, touch objects to feel their shape or texture, and walk around to understand a space. These movements help the brain form predictions about what it expects to sense next. It then compares these predictions to the actual input. If the prediction is correct, the brain strengthens its understanding. If not, it adjusts and learns.

This ongoing feedback loop between sensing and acting is not only important for learning. It also supports more advanced abilities like planning and imagination, which often involve mentally simulating actions before we carry them out.

2.2.6 *The Thousand Brains Framework*

The Thousand Brains Theory proposes that the neocortex is made up of thousands of small units called cortical columns. Each column acts like a mini-processor, independently learning about objects or concepts based on limited sensory input, such as a small area of the skin or a portion of the visual field. Although each column processes only part of the whole, they work together by sharing information. Through a kind of voting process, they combine their individual models to create a more complete and accurate understanding of what is being sensed.

Instead of depending on a single source, perception arises from the combined input of many cortical columns. For example, when identifying an object, visual and tactile columns each generate their own guesses based on what they sense. By combining these inputs, the brain creates a single, unified perception. This distributed structure makes the brain more resilient because damage to a few columns does not cause major problems. It also allows the brain to integrate information from multiple senses into rich, abstract concepts.

A simple action like drinking tea shows how this system works. As you reach for the cup, different columns handle touch, sight, and movement. They make predictions about what should happen next and adjust your grip as needed. Each column provides only part of the picture, but together they guide smooth, coordinated behavior.

2.3 Neocortex Cortical Layers

2.3.1 Hierarchical and Synchronization

The brain processes sensory inputs, like vision, sound, and touch, through a hierarchical system that transforms raw signals into abstract understanding. Lower levels detect simple features (e.g., edges, tones), while higher levels integrate them into stable patterns like objects or words.

Crucially, the brain combines inputs across senses through cross-modal integration. For example, in a noisy cafe, hearing someone say "I'm hungry" involves auditory cues (pitch, phonemes) and visual signals (lip movements, expressions). These are fused into a unified, abstract meaning. This multisensory integration enables robust perception even when individual inputs are noisy or incomplete.

A key feature of the brain's processing system is its ability to make predictions. This depends on two important mechanisms: cross-modal integration and top-down processing. Cross-modal integration allows the brain to combine information from different senses to create a fuller understanding of a situation. Top-down processing uses prior knowledge, past experiences, and context to anticipate what is likely to happen next. Working together, these mechanisms help the brain interpret the world quickly and accurately, even when information is incomplete or unclear.

The neocortex operates using a general learning framework built on four core principles that support recognition, prediction, and abstraction across all forms of sensory input, not just sequences.

- **Invariant pattern representation**: The brain can recognize objects or concepts despite variations in appearance, sound, or context. For example, it can identify different dog breeds as "dogs" based on shared characteristics.
- **Pattern learning across time and space**: Information is learned not just as static snapshots but as evolving patterns, allowing the brain to anticipate what comes next or how things relate spatially and temporally.
- **Completion from partial input**: Even when only part of a pattern is available, the brain can retrieve the whole based on previous experiences, such as recognizing a familiar face in poor lighting.
- **Hierarchical organization**: Simple features are processed at lower levels and gradually combined into more complex and abstract representations at higher levels.

This system is implemented in the layered structure of the neocortex, which contains six interconnected layers. Lower layers receive raw sensory data, such as visual edges or sound frequencies, and higher layers integrate these signals into meaningful forms such as words, scenes, or objects. Figure 2.2 illustrates how information flows through this hierarchical structure.

This architecture enables both **bottom-up** and **top-down** information flow:

- **Bottom-up processing**: Raw sensory input is passed through successive layers, each extracting increasingly abstract features. For example, in language, the pro-

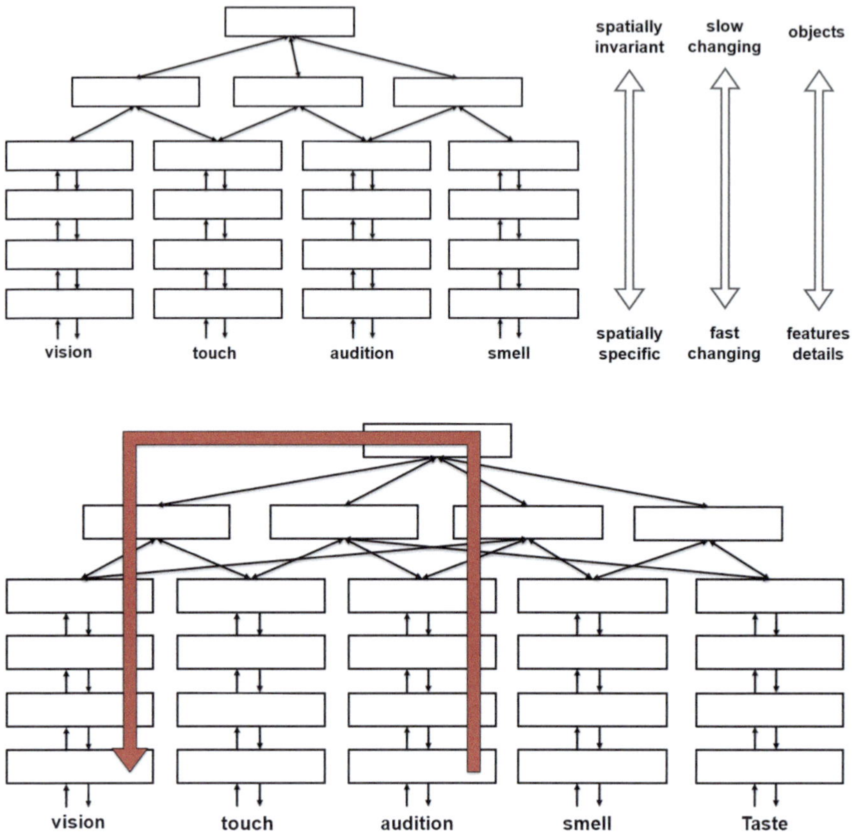

Fig. 2.2 The structure of cortical hierarchy and information flow

cess moves from phonemes to words to meanings. In vision, it progresses from visual lines to objects.

- **Top-down processing**: Higher layers send predictions and expectations back to lower layers, guiding interpretation and attention. This helps the brain manage ambiguous or noisy input by incorporating context and prior experience. Top-down processing also supports imagination, mental simulation, and goal-driven behavior. It enables us to visualize a scene, plan an action, or predict an outcome, activating sensory or motor areas even in the absence of external input.

This bidirectional interaction helps the brain align new input with existing knowledge. When incoming information matches what is expected, recognition happens quickly and efficiently. However, when the input does not match expectations, the brain becomes alert. This signals that something is unfamiliar or surprising. In response, the brain updates its internal model by adjusting existing patterns or forming new ones. This ongoing cycle of matching, predicting, detecting surprises, and adapting is at the core of how the brain learns and makes sense of the world.

2.3.2 Cortical Layers

The neocortex, the outermost layer of the brain responsible for higher-order cognition, consists of six distinct but highly interconnected layers. These layers collaborate to process sensory input, form interpretations, and generate appropriate responses. Each layer contributes a specific function, yet together they form a dynamic system that supports perception, learning, and behavior.

At the surface, Layer 1 (L1) primarily facilitates horizontal integration. It enables communication across nearby cortical regions and columns, allowing the brain to merge contextual cues from different senses and spatial areas. Importantly, L1 also receives top-down feedback from higher cortical regions, integrating expectations, attention, and prior knowledge into early sensory processing.

Layer 4 (L4) acts as the main entry point for external sensory input. Stimuli such as light, sound, or touch are relayed via the thalamus and arrive in L4 of the relevant sensory cortex. This layer specializes in detecting basic features such as edges, tones, and vibrations, and contains stellate neurons that distribute this raw data to the layers above for further analysis.

Sitting just above, Layers 2 and 3 (L2/3), known as the supragranular layers, play a vital role in pattern recognition and association. Layer 2 mainly handles local processing within a cortical column, while Layer 3 transmits long-range signals to other cortical regions, especially association areas. These layers integrate inputs, identify patterns, and build meaningful representations that connect perception to cognition. They also serve as the primary source of feedforward connections to higher-order regions, advancing the information through the cortical hierarchy.

Once processed, information flows downward to Layers 5 and 6, which handle output generation and feedback control. Layer 5 is crucial for initiating actions: it sends motor commands to the brainstem and spinal cord via the corticospinal tract, and also projects to subcortical structures like the basal ganglia and superior colliculus to coordinate movement sequences. In contrast, Layer 6, located at the base of the neocortex, focuses on modulation and regulation. It projects back to the thalamus, specifically the thalamic reticular nucleus, helping adjust the gain and timing of sensory signals. It also contributes to attentional filtering, ensuring that relevant information is prioritized based on the brain's current goals.

In addition to this vertical flow of information, the neocortex features several feedback loops that refine perception in real time. For instance, the corticothalamic loop allows Layer 6 to influence how sensory data is relayed from the thalamus. Simultaneously, top-down projections from regions like the prefrontal cortex reach Layer 1 in primary sensory areas, modulating pyramidal neurons' activity deeper in the cortex. This enables the brain to align sensory input with expectations, past experiences, and attentional focus.

Finally, the neocortex helps the left and right sides of the brain work together by supporting interhemispheric integration. This is done through neurons in Layers 2/3 and 5, which send signals across the corpus callosum, a thick band of nerve fibers that connects the two hemispheres. These signals link both matching areas, such as the left and right visual cortex, and different areas, such as the frontal and parietal lobes.

This communication allows the brain to combine information from both sides of the body. It supports tasks like seeing with both eyes, moving both hands together, and connecting language from the left hemisphere with emotional or visual context from the right.

When this connection is damaged, as seen in split-brain patients, thinking and perception can become disconnected. For example, they may see an object with their left eye but be unable to name it because the information does not reach the language center on the left side. Although other brain pathways like the anterior commissure can offer limited support, the corpus callosum plays a central role in creating a unified view of the world.

The structure and function of cortical layers play a crucial role in how the brain processes information. To offer a clearer conceptual understanding, Fig. 2.3 presents a simplified representation of neocortical activity, illustrating how information is transmitted and interpreted across six interconnected layers. When the brain receives a new sensory input, such as the feature represented by "B," the thalamus relays this signal to Layer 4 (L4) of the neocortex, the primary input layer. In L4, specialized neurons detect basic features of the input and distribute them across multiple cortical columns, each of which has previously learned representations.

In this case, several columns activate in response to "B." The column labeled "B" becomes most strongly activated, as it best matches the incoming pattern. Other nearby columns (e.g., "R" or "P") may show partial activation if they share features with the input, reflecting the brain's ability to generalize.

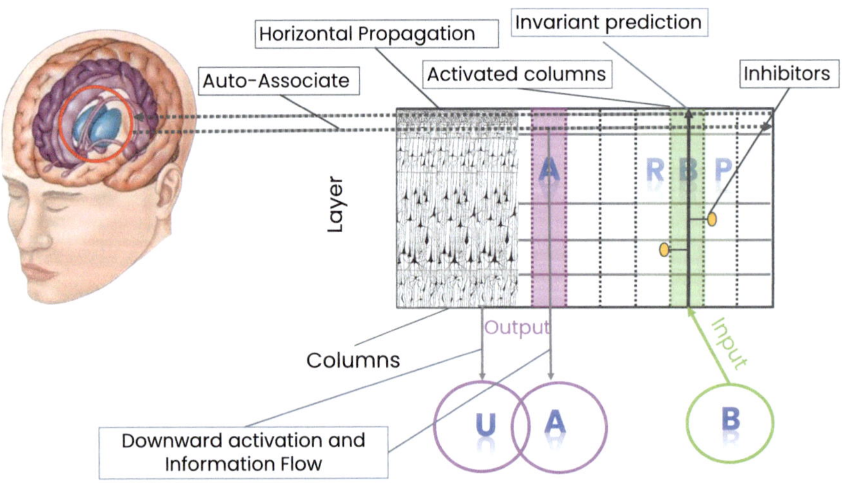

Fig. 2.3 Transmission and processing of information in cortical layers

To refine recognition, inhibitory neurons (depicted as yellow circles) suppress competing columns, ensuring that only the most accurate one, "B," remains active. This winner-takes-all mechanism enhances precision.

Once the column representing "B" is activated, a signal is sent to the thalamus, which draws on learned sequences and contextual memory to predict what should come next. Based on past experience, the thalamus identifies "A" and "U" as likely successors and pre-activates those representations in the neocortex. This pre-activation primes the system for rapid, seamless transitions in perception, thought, or behavior.

However, if the actual next input does not match the predicted outputs, such as something other than "A" or "U" appearing, the brain registers a mismatch or prediction error. This discrepancy triggers an exception signal, which is routed back through the thalamocortical loop to redirect attention and update the current context. Such unexpected inputs prompt the brain to re-evaluate its assumptions, learn from the new pattern, or respond more cautiously.

2.3.3 *Reflection on Learning Mechanism*

This section consolidates the concepts presented in the preceding sections to reflect how human thinking is modeled, as illustrated in Fig. 2.4.

- The process begins when sensory information from the five human senses is transmitted to the thalamus, which then relays the signals to the lower layers of the neocortex. From there, the information flows upward through the hierarchical lay-

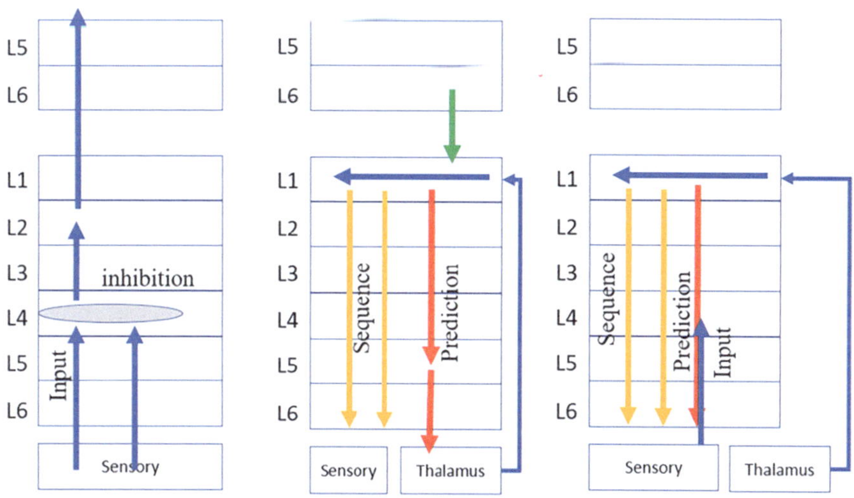

Fig. 2.4 The concept of how humans think

ers of the neocortex. During this process, both excitatory and inhibitory signals are involved. As the input is processed, one column or representation is selectively activated while competing columns are inhibited. This mechanism ensures clarity and precision in interpretation.
- In each layer of the hierarchy, long-term memory is used to recognize patterns and make predictions. When a pattern in the current layer matches the input, a prediction is generated and sent downward to influence lower layers. This feedback is shown in green in Fig. 2.4. The lower layer then uses the invariant representation from the upper layer to generate more specific predictions (red arrows). Additionally, the thalamus contributes by projecting signals back to Layer 1 to help form and reinforce sequence memory (yellow arrows).
- If no valid prediction can be made at a given level, the input is forwarded to the next higher layer in the hierarchy. At the top of this hierarchy is the hippocampus, which temporarily stores unpredicted or novel information in short-term memory. If the input persists and becomes familiar through repeated exposure, it is gradually consolidated into the neocortex as new long-term memory. This process results in the formation of new neural connections that are integrated with existing representations, allowing the brain to continuously learn and adapt.
- Finally, when new inputs are received, they are compared with the brain's predictions. If the input matches the expected pattern, processing continues without disruption. However, if the input does not match, the discrepancy is treated as an unexpected event, and the signal is escalated to higher levels of the hierarchy for further evaluation.

2.4 Implications for Machine Intelligence

Building on insights from both the previous and current chapters, the structure and function of the neocortex have been foundational to the development of neuroscience-inspired models. One such model is Hierarchical Temporal Memory (HTM), a theory introduced by Jeff Hawkins in his 2004 book On Intelligence. HTM is grounded in the neuroscience understanding of the neocortex and reflects its hierarchical and modular architecture.

Unlike modern AI techniques such as deep learning, which rely on large numbers of mathematically defined neurons, HTM emphasizes a biologically plausible approach. It models the neocortex as a uniform hierarchy composed of cortical columns, each containing multiple neural cells. These cells are interconnected through dendrites and synapses: proximal connections handle direct input and feed-forward signals, while distal connections are involved in contextual prediction.

HTM employs Hebbian learning, in which synaptic connections are strengthened when cells are activated simultaneously and weakened when they are not. Through this mechanism, HTM systems can learn temporal patterns and make predictions based on contextual input. Importantly, HTM is specifically designed to approximate the functioning of the neocortex, drawing on the work of Hawkins, contributions from other neuroscientists, and the current body of neuroscience research.

2.4.1 Structure

The area in the neocortex is made up of cortical regions. within the regions comprising cortical columns or smaller, macrocolumns, that are a group of minicolumns. Each minicolumn is a group of neurons. A minicolumn is believed to be the basic building block of cortical columns and plays a key role in information processing in the neocortex. Inside, the mini-column contains six layers (L1-L6) that perform different tasks. HTM replicates these working only L2 and L3 layers. Figure 2.5 shows the difference between the cortical column in the neocortex and HTM.

In contrast to deep learning that uses mathematically derived neurons for back-propagation, HTM more closely resembles a biological neuron as it operates independently and learns through strengthening and weakening connections via Hebbian-like learning. Figure 2.6 shows a biological neuron on the left, a deep learning neuron in the middle, and an HTM neuron on the right. HTM cell contains three inputs, *feedforward* to receive input from the below layer or sensory, *context* to connect

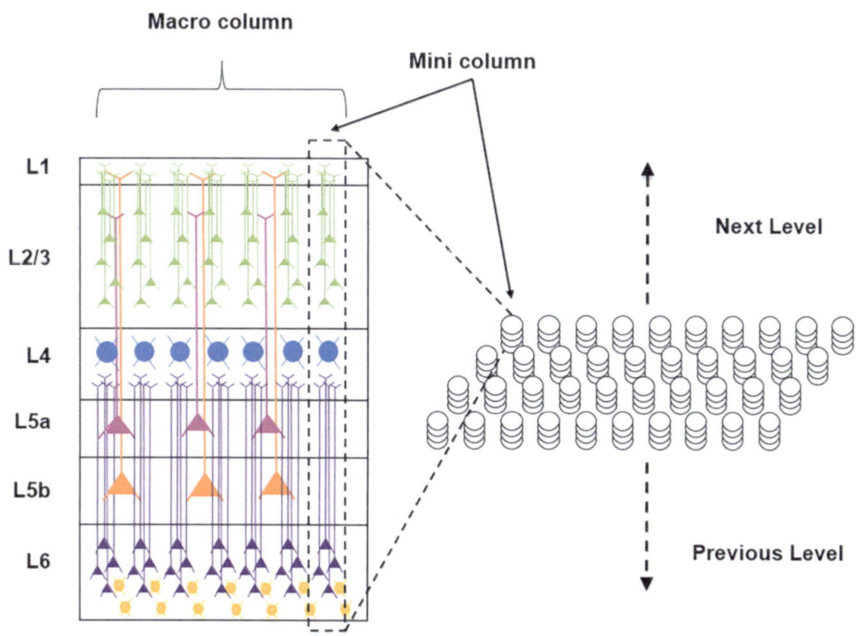

Fig. 2.5 Visualization of cortical minicolumns within a macrocolumn in the neocortex vs HTM region and its columnar structure and cells

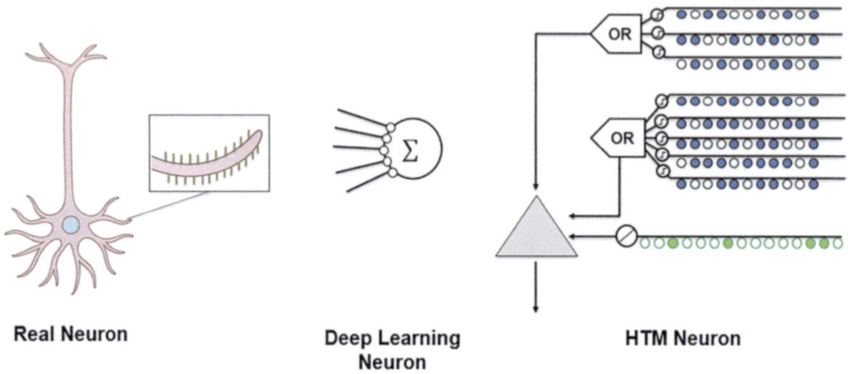

Fig. 2.6 Comparison between biological, HTM and deep learning neurons

and predict to its neighbors, and *feedback* to get and predict from other layers in the hierarchy. Furthermore, HTM is a common algorithm, is capable of handling various inputs, but these inputs must be transformed into the appropriate format via an encoder before being processed by HTM. In other words, all inputs use the same format.

2.4.2 Spare Distributed Representation (SDRs)

The concept of Sparse Distributed Representations (SDRs) [3] is a fundamental component of Hierarchical Temporal Memory (HTM), used for efficient information storage and transmission. SDRs represent information as binary vectors composed of "0"s and "1"s, where only a small percentage of the bits are active at any given time. This sparse activation mirrors the brain's strategy of minimizing neural activity to conserve energy and reduce noise in inference.

HTM utilizes SDRs to recognize both spatial and temporal patterns through its Spatial Pooler and Temporal Pooling mechanisms. The concept of SDRs will be revisited in greater detail in Chap. 4.

2.4.3 Encoder

HTM is capable of processing any type of input using a unified algorithmic framework. This flexibility is made possible through the use of encoders, which transform diverse input signals into Sparse Distributed Representations (SDRs) before they are processed by the HTM system. Designing an effective encoder, however, is a non-trivial task, as it must preserve essential features of the input to ensure meaningful and accurate representation. The choice of encoder plays a crucial role in determining the overall performance of the model.

This encoding process mirrors the biological pathway in which visual information is transmitted from the retina to the neocortex, where it is interpreted and learned. The design of encoders that generate Sparse Distributed Representations (SDRs) follows a set of core principles, as described in [4]:

- SDRs of semantically similar data should have overlapping active bits.
- The SDR generated from a particular input should remain consistent and not change over time.
- The number of bits in the output should be consistent for all inputs.
- The sparsity of the output should be similar for all inputs, and it should have sufficient one-bits to handle subsampling and noise.

Examples of encoding can be found in [5]. However, encoding in NLP is not mentioned but refers to cortical.io, which is a commercial api encoding to SDRs and only provides a concept in [6]. Shah et al. [7] uses HTM for document categorization. It uses TF-IDF, finds Latent Semantic Indexing (LSI), and encodes numerical features into SDRs. Rydholm [8] is anomaly detection in system logs that uses GloVe word embeddings [9] and encodes the numbers into SDRs. As aforementioned, NLP in HTM commonly uses encoding numbers into SDRs.

2.4.4 *Learning*

Hierarchical Temporal Memory (HTM) is designed to work with streaming data, similar to how the neocortex operates. This approach, known as online learning, contrasts with the batch learning methods used in traditional machine learning. HTM has a two-step mechanism consisting of a spatial pooler (SP) for pattern recognition and a temporal memory (TM) for sequence memory. SP receives SDRs from the encoder and uses Hebbian learning to find important features and set them into output SDRs. The output SDRs send up to TM to learn sequences and perform prediction. The system will learn and minimize the error between the predicted and input values. Thus, the system continuously adapts to the environment. Figure 2.7 shows an end-end HTM flow for a classification problem that passes an encoder, SP, and TM.

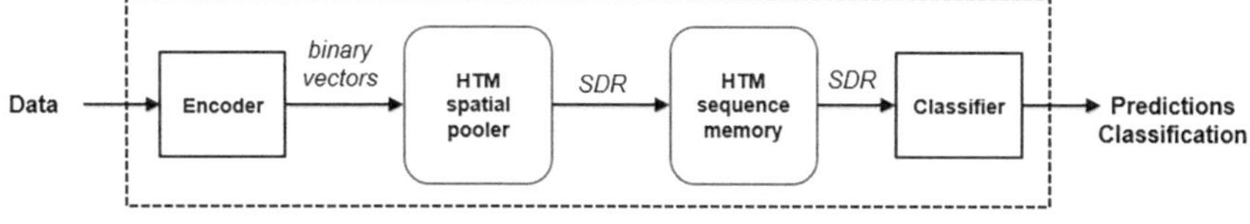

Fig. 2.7 An HTM system consists of an encoder, the HTM spatial pooler, the HTM sequence memory, and an SDR classifier

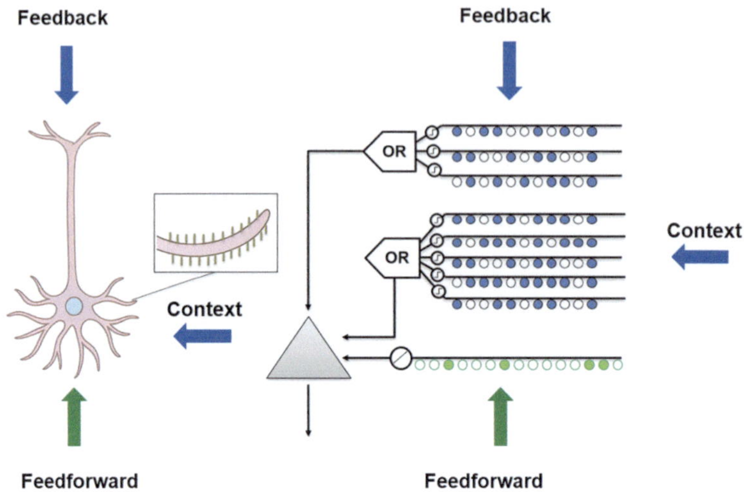

Fig. 2.8 On the left is a neocortical pyramidal neuron and on the right is an HTM model neuron. The green color in the feedforward represents SP, while the blue in the context and feedback represents TM

Figure 2.8 illustrates the connections and learning processes of a neuron in HTM, drawing a comparison with an actual neural cell. The green color in the feedforward represents Spatial Pooling (SP), while the blue color in the context and feedback signifies Temporal Memory (TM).

2.4.4.1 Spatial Pooler

The spatial pooler (SP) in HTM is made up of columns, which can be considered as nodes that each cover a receptive field over the input. During initialization, each column establishes random connections to a portion of bits in the input space within its receptive field, resulting in a distinct potential pool. This can result in overlapping receptive fields when the potential radius is large.

The strength of each connection in SP is described using a permanence value that ranges from 0 to 1. During learning, the permanence value is adjusted based on the activity of the input bit.

When the permanence value of a connection surpasses a stimulus threshold, the connection is considered "active". The overlap score is the number of active connections for a particular column, and if the overlap score is sufficiently high, the column is activated. By setting a minimum overlap score requirement, the system can effectively filter out input noise. Finally, the Spatial Pooler (SP) output only considers the top n columns with the highest overlap scores, where n is determined by the desired sparsity level. These selected columns are then allowed to learn and adjust their permanence values.

The active columns in SP are the only ones that can learn and adjust their permanence values. This can lead to a few columns dominating the output with a high active-duty cycle since initialization, which measures the frequency of a column's activity. To balance the dominance of columns, HTM uses boosting, which increases the overlap score of underperforming columns and decreases the overlap score of overperforming columns. This allows more columns to learn and contribute to the output with finer granularity. However, if not used carefully, boosting can cause instability in the output. In addition, columns can be selected based on their local neighborhood rather than their global overlap score, called *topology*.

2.4.4.2 Temporal Memory

To learn sequences and make predictions, the Temporal Memory (TM) algorithm is utilized by establishing connections between cells that were active in the past. By using these connections, the cells can anticipate their future activity. This ability enables the cells to store and retrieve sequences collectively, which in turn facilitates the prediction of future events. The memory of the system is distributed across its cells, making it robust to noise and errors. Therefore, if some individual cells fail, the system as a whole is usually not adversely affected.

TM, or Temporal Memory, consists of columns that are the output of the spatial pooler. These columns contain neuron cells, and the number of cells in a column determines the capacity for context representation. Cells within a column can form connections to other cells in other columns through segments. Each segment connects to other cells via synapses.

To represent a pattern in multiple contexts, the Temporal Memory (TM) algorithm uses a subset of cells per column, typically only one cell per column. The representation of a pattern is modified to include state or context information from previous time steps. By having 12 cells per column and 1000 active columns in the Spatial Pooler (SP) output, the TM algorithm can represent the same input in different ways, with up to 12^{1000} unique representations, by activating different cells in different contexts.

The Temporal Memory (TM) algorithm consists of two phases. In the first phase, if any predictive cells exist within an active column, they are designated as active. If there are no predictive cells, then the column "bursts" by activating all its cells. Conversely, if the column still does not contain any active cells, it remains inactive. During the bursting phase, the TM algorithm examines the segments that contain any active synapses. If no active segments are present, a new segment is created on the cell with the fewest segments. For each examined segment, the algorithm enhances the permanence of every active synapse, reduces the permanence of every inactive synapse, and creates new synapses to cells that were previously active. The algorithm also penalizes segments that prompted cells to enter a predictive state but ultimately did not become active.

In the second phase of TM, a prediction is generated by putting cells into a predictive state. The algorithm counts the number of synapses connected to active

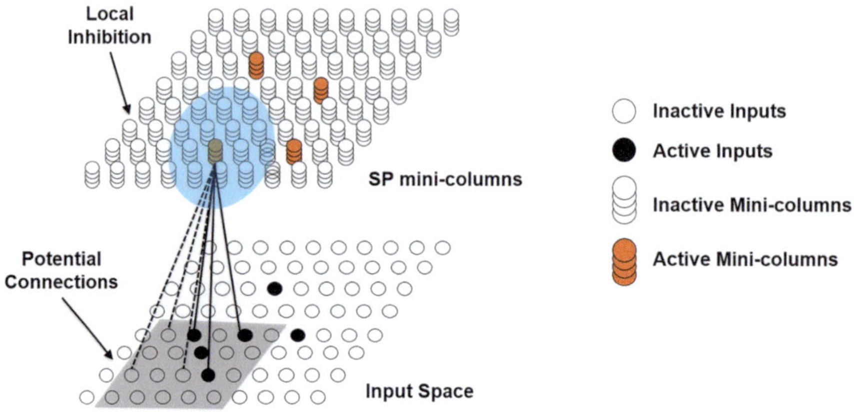

Fig. 2.9 Learning by using SP and TM in HTM

cells for each segment on each cell. If the count exceeds an activation threshold, the segment is marked as active, and the connected cells enter the predictive state. There are three possible states that a cell can be in.

- **Active**: if either the column is bursting or the cell was previously in a predictive state.
- **Predictive**: if a connected segment is active.
- **Inactive**: neither active nor predictive state.

The learning ability of the system can be controlled by setting the number of cells and adjusting permanence values. Setting the permanence decrement value to 0 will prevent forgetting. To only show patterns in the current and previous context, the number of cells can be set to 2.

Figure 2.9 illustrates the learning process in HTM using Spatial Pooling (SP) and Temporal Memory (TM). SP transforms inputs into Sparse Distributed Representations (SDRs). Within SP, each mini-column forms connections to a subset of the input space, and a local inhibition mechanism ensures that only a small number of mini-columns within a specific radius are active. The adjustment of synaptic permanences is guided by the Hebbian rule, reinforcing active connections and penalizing inactive ones.

2.5 Conclusion

This chapter offered a focused exploration of the neocortex, the brain's most advanced region responsible for higher-order cognition. Each section was structured to build a layered understanding, beginning with the biological foundation and culminating in its relevance to machine learning. Section 2.1 introduced the anatomical structure and four core attributes of the neocortex: uniformity, invariant representation, hierarchy and auto-associative memory. These structural and functional principles are crucial, as they enable the neocortex to process complex, multi-modal information and provide a blueprint for general-purpose learning.

Following this, Sect. 2.2 presents the fundamental ideas proposed by Jeff Hawkins, particularly the view that intelligence stems from common computational mechanisms used throughout the neocortex. This section bridges neuroscience with computational theory by introducing concepts such as hierarchical learning, temporal memory, prediction, and sensory-motor integration, each reinforcing and expanding upon the attributes introduced earlier.

Next, Sect. 2.3 dives into the layered cortical structure and explains how different layers interact to perform both bottom-up and top-down processing. This section operationalizes the theoretical ideas introduced earlier by showing how the brain's physical structure supports memory, prediction, and learning.

Finally, Sect. 2.4 translates these previous biological and theoretical insights into a practical model: Hierarchical Temporal Memory (HTM). This section demonstrates how the neocortex's architecture and functionality inspire computational systems that mimic human-like learning. It shows that HTM incorporates sparse representations, online learning, and temporal sequence modeling, reflecting key principles introduced throughout the chapter.

This understanding of the neocortex, both biological and computational, leads directly into Chap. 3, which explores key approaches in Artificial Intelligence (AI). The next chapter reviews major AI methods, highlighting their strengths, limitations, and how they differ. It also identifies gaps in current technologies and opportunities for improvement by connecting ideas from both AI and neuroscience. These insights help build a foundation for developing more advanced and brain-inspired intelligent systems.

References

1. Hawkins, J., Blakeslee, S.: On Intelligence. Times Books, New York (2004)
2. Mountcastle, V.B.: An organizing principle for cerebral function: the unit module and the distributed system. In: Edelman, G.M., Mountcastle, V.B. (eds.) The Mindful Brain, pp. 7–50. MIT Press, Cambridge, MA (1978)
3. Ahmad, S., Hawkins, J.: Properties of Sparse Distributed Representations and their Application to Hierarchical Temporal Memory (2015). arXiv:1503.07469. https://arxiv.org/abs/1503.07469
4. Hawkins, J., Ahmad, S., Purdy, S., Lavin, A.: Biological and Machine Intelligence (BAMI). Initial online release 0.4 (2016). https://numenta.com/resources/biological-and-machine-intelligence/. Accessed 4 May 2025
5. Purdy, S.: Encoding Data for HTM Systems (2016). arXiv:1602.05925. https://arxiv.org/abs/1602.05925
6. Webber, F.D.S.: Semantic Folding Theory and its Application in Semantic Fingerprinting (2015). arXiv:1511.08855. https://arxiv.org/abs/1511.08855

7. Shah, D., Ghate, P., Paranjape, M., Kumar, A.: Application of hierarchical temporal memory theory for document categorization. In: 2017 IEEE SmartWorld, Ubiquitous Intelligence and Computing, Advanced and Trusted Computing, Scalable Computing and Communications, Cloud and Big Data Computing, Internet of People and Smart City Innovation, pp. 1–6. IEEE, San Francisco, CA (2017). https://doi.org/10.1109/UIC-ATC.2017.8397473
8. Rydholm, G.: Using Neurobiological Frameworks for Anomaly Detection in System Log Streams. Master's Thesis, KTH Royal Institute of Technology, Stockholm, Sweden (2018)
9. Pennington, J., Socher, R., Manning, C.: GloVe: global vectors for word representation. In: Proceedings of the 2014 Conference on Empirical Methods in Natural Language Processing (EMNLP), pp. 1532–1543. Association for Computational Linguistics, Doha, Qatar (2014). https://doi.org/10.3115/v1/D14-1162

Chapter 3
The Brain and Methods of ML

Building on the biological foundations of intelligence discussed in Chap. 1, particularly the structure and learning mechanisms of the neocortex in Chap. 2, in this chapter transitions to the field of Artificial Intelligence (AI). It examines the rationale behind key AI methods, how each one operates, the types of problems they are best suited to solve, and where their limitations lie.

This chapter serves two critical purposes. First, it provides a comprehensive survey of key AI methodologies, from early rule-based systems to modern machine learning techniques. Second, it offers a comparative analysis of their strengths, limitations, and the degree to which they align with biological principles. By integrating insights from neuroscience and AI research, the chapter lays the groundwork for developing more brain-inspired intelligent systems in subsequent chapters.

In this chapter begins with an overview of Artificial Intelligence (Sect. 3.1), introducing its interdisciplinary foundations in philosophy, mathematics, economics, and neuroscience. It explains how AI agents interact with different types of environments, including deterministic, stochastic, episodic, and dynamic, and categorizes them based on observational scope and control. Section 3.2 explores symbolic approaches to AI through rule-based and knowledge-based systems, using propositional and first-order logic to enable inference and reasoning. Section 3.3 continues with search algorithms, dividing them into uninformed (blind) and informed (heuristic) methods, primarily applicable in well-defined, static environments.

The next part of the chapter focuses on machine learning. Section 3.4 introduces supervised, unsupervised, and reinforcement learning, covering widely used algorithms like decision trees, ensemble models, k-nearest neighbors, and support vector machines. Section 3.5 delves into deep learning, examining architectures such as feedforward and convolutional neural networks. It emphasizes advantages like feature learning and scalability while also discussing challenges such as computational cost and black-box behavior. Section 3.6 addresses sequential learning methods for temporal data, including ARIMA, RNNs, LSTMs, GRUs, and Seq2Seq models,

T. Soisoonthorn and H. Unger, *A Brain-Inspired Approach to Natural Language Processing*, Studies in Computational Intelligence 1225,
https://doi.org/10.1007/978-3-032-00014-9_3

emphasizing their role in tasks that require memory and time-aware processing, as well as their inherent limitations.

Section 3.7 presents the Transformer architecture and the development of Large Language Models (LLMs), detailing how these advanced models have transformed sequence modeling and language understanding through mechanisms like self-attention and multi-head attention. Section 3.8 introduces reinforcement learning as a dynamic framework in which agents adapt their behavior through trial and error, guided by feedback in the form of rewards. This approach enables decision-making in uncertain and changing environments by learning optimal policies over time. Section 3.9 offers a critical analysis of current AI methodologies, highlighting persistent challenges such as vulnerability to noise, limited contextual understanding, and difficulties in generalizing across different tasks. Section 3.10 serves as the conclusion of the chapter.

3.1 Artificial Intelligence

Artificial Intelligence (AI) refers to any device or intelligent agent that receives percepts from the environment over time, adapt to its behavior, and performs actions to accomplish its goals, as Fig. 3.1. In some definitions, AI cites the simulation of human intelligence in a machine that can do the planning, learning, reasoning, knowledge representation, natural language processing, perception, and the ability to do intelligent behavior. It has been one of the most significant emerging technologies in today's world. Now and then, most things in the world are adapting to AI, such as agriculture, cybersecurity, education, finance, military, healthcare, law, and transportation.

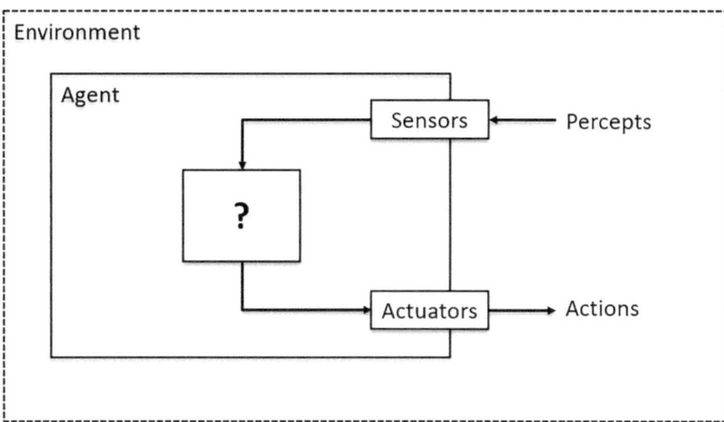

Fig. 3.1 AI agent that perceives its environment through sensors and actions upon the environment through actuators

As Homo sapiens, our intelligence distinguishes us from all other animals. For thousands of years, we have tried to understand how we think and do complicate things. Our intention is not only to understand how we think but also to build AI. The modern AI techniques ideas are contributed from various fields as the following.

- **Philosophy** human uses relation and logic to reasoning and conclusion. To emulate how humans think, Aristotle (384-322 BCE), at first, created *syllogisms* for proper reasoning, which allowed them to generate conclusions and induction logically. It has been developed until now and is called *logic*.
- **Mathematics** in formal logic is used to draw valid reasoning and conclusion. While probability and statistics use to handle uncertain information.
- **Economics** is used to make a decision to achieve its goal. Decision theory combines probability and utility theory for individual decisions under uncertainty. Alternatively, Game theory states that one player's actions can significantly affect others' utility.
- **Neuroscience** is a field to study how the brain works. The studies include neurons, memory, motor, emotion, sensory, or processing connections for each part.
- **Control Theory** can modify their behavior in response to environmental change by maximizing a cost function over time.

One consideration of designing AI agent depends on the environment in which the agent interacts, which causes different solutions. We can categorize and design AI agents in different environments as follows.

- **Fully Observation or Partially Observation**: If an agent can access all complete states of the environment at any time, then the environment is fully observed. Otherwise, partial observation.
- **Deterministic or Non-deterministic**: In real-world situations, the environment is non-deterministic as the agent cannot predict an uncertain environment, or the environment is partially observable. However, if the environment is completely determined, has no randomness, and always produces the same output, then it is deterministic (Fig. 3.2).
- **Episodic or Sequential**: In Episodic, the agent gets a percept and acts. The action does not depend on the previous episodes. Classification, for example. Otherwise, it is sequential.
- **Discrete or Continuous**: If the percepts or actions of an agent are a range of continuous values and smoothly over time, then it is continuous. Speed or controlling wheel, for instance. Chess or Go is discrete.
- **Known or Unknown**: From known, an agent can predict outcomes from all actions. It needs to learn how to take good action if it is unknown.
- **Static or Dynamic**: If the environment can change, it is dynamic over time. Otherwise, it is static.
- **Single-Agent or Multiagent**: Multiagent is used to maximize overall performance. For instance, cars with autopilot avoid collisions to maximize the performance of all agents. Single-Agent performs individually to maximize its performance.

The first solution of AI is using *Rule-base* (RB) by using knowledge from humans and convert into rules or if-then-else statements. It can help if there is a danger of errors, speed, and no data. However, it cannot make inferences or produce new information that differs from the rules. Another AI system close to Rule-base is called *Knowledge-base* (KB). KB can infer or produce new information logically. KB is presented in Sect. 3.2. Nevertheless, suppose some problems have large solution spaces and are a completely known model that needs to find the optimal solution effectively. In that case, a search algorithm can be used to solve the problems discussed in Sect. 3.3.

But, sometimes a problem has only data; one approach is using data to learn and improve behavior, called *machine learning*, which can be found in Sect. 3.4. Furthermore, this section mentions the state-of-the-art methods such as Random Forest, K-Nearest neighbors (kNN), Support Vector Machine (SVM). Finally, deep learning is separated into another Sect. 3.5 as it is a success of modern AI.

Nonetheless, if the environment is temporal, meaning it consists of sequences of data rather than isolated events, using common machine learning techniques does not provide good results, even if they can analyze fixed-length information. To address this challenge, traditional methods such as Exponential Smoothing and Auto-Regressive Integrated Moving Average (ARIMA), as well as modern approaches like Recurrent Neural Networks (RNNs), Long Short-Term Memory (LSTM) networks, Gated Recurrent Units (GRUs), and Sequence-to-Sequence (Seq2Seq) learning, are introduced in Sect. 3.6. However, RNN-based models face several limitations. They have difficulty handling long-range dependencies, cannot capture semantic information from long sequences, and process inputs sequentially, which leads to slow training and inference. To overcome these challenges, the Transformer model was introduced. It can capture long-range dependencies and process data in parallel, making it one of the most successful models today, as discussed in Sect. 3.7.

In Sects. 3.4 through 3.7, the machine learning learns by passively observing data only. But if no training data or the training data might not be enough to train in a situation due to many possible actions and states. For example, we can train a machine to learn Chess from several million grandmaster games for each sequence of positions and moves. Nevertheless, training data only contains 10^8 few examples from grandmaster games compared to 10^{40} for all possible Chess positions. Reinforcement learning (RL) copes with this problem by performing trial and error to determine the best action to maximize cumulative reward. The algorithm itself generates the model. It is similar to the search algorithm, but the task is an unknown model. RL can be found in Sect. 3.8. Last but not least, Sect. 3.9 critically evaluates the state of the art and current approaches, highlighting both their advantages and disadvantages, as well as identifying areas for improvement and gap analysis.

Each mentioned method can be summarized and categorized into four categories, shown in Fig. 3.3 and follows.

1. **Knowledge representation** the model uses knowledge to create a model from humans manually. There are two current approaches as the following.

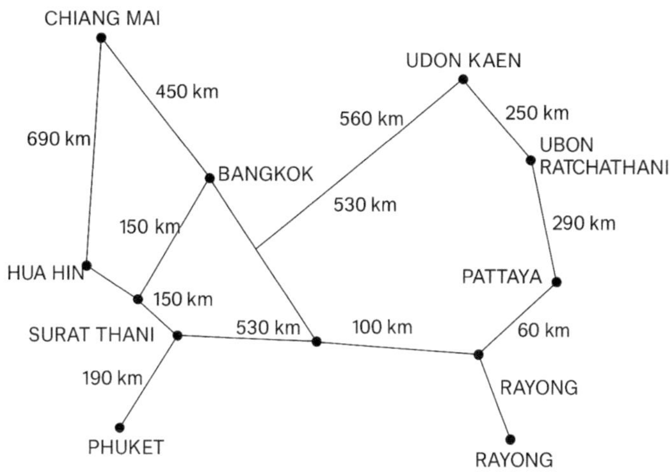

Fig. 3.2 Vehicle routing problem with fully observable and deterministic environment

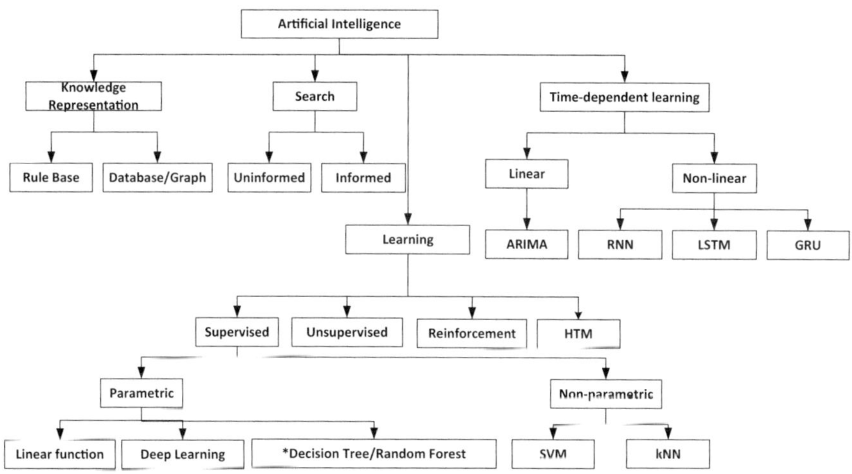

Fig. 3.3 The types of AI

a. Rule base is hard code into the model.
b. Database or Graph base, the model is created from knowledge and inserted into a database or graph database in a specified format. New knowledge can be inferred from the current information. However, humans need to manually input new knowledge or things from the environment.

2. **Search** is finding solutions from state spaces, but the environment must be deterministic.
   a. Uninformed search, the method looks through search space for all possible.
   b. Informed search; the method uses some existing information to help in finding a good solution. It might not search all possible spaces, but it finds more efficient ways to traverse less in the search spaces and find a good solution in a reasonable time.
3. **Machine Learning** is the process of obtaining new knowledge from experiences and keeping it in the model. It composes of three classes.
   a. Supervised learning is learning from training data by teaching output and input. For instance, classification and regression. The methods include Decision tree, Random forest. K-nearest neighbors and support vector machine for non-parametric model as well as deep learning that is inspired by neuron connections in the brain. Feed-forward neural networks and Convolution Neural networks, for example.
   b. Unsupervised learning is used to explore interesting structures in data without teaching or labeling output.
   c. Reinforcement learning performs actions as a trial and error to learn what behaviors can gain a maximum reward. Hence, the agent can be adapted as the environment changes. It resembles a control system that is controlled with parameters that vary and uncertainty.
   d. HTM distinguishes itself from other types of AI by its ability to perform online learning and adapt to its environment through replicating the functions of the neocortex.
4. **Time-dependent learning** involves learning the correlation between the current input and previous inputs or states. In other words, the next step $x(t)$ depends on some previous steps $x(t-k)$. Linear methods include ARIMA, while non-linear methods are based on deep learning models such as Recurrent Neural Networks (RNNs), Long Short-Term Memory (LSTM) networks, Gated Recurrent Units (GRUs), and Sequence-to-Sequence (Seq2Seq) learning. LSTM and GRU incorporate memory mechanisms that allow information to be forgotten, remembered, and read as needed. The Transformer model, which processes input data in parallel rather than sequentially, is also a powerful approach for modeling time-dependent data.

3.2 Knowledge Representation

Knowledge-base (KB) or Knowledge Representation is based on philosophy and logic that was invented by Aristotle (BCE) by creating syllogisms for proper reasoning, which allowed to generate conclusions and induction logically. KB is transformed into mathematical logic in the mid-19th. Propositional logic (PL)

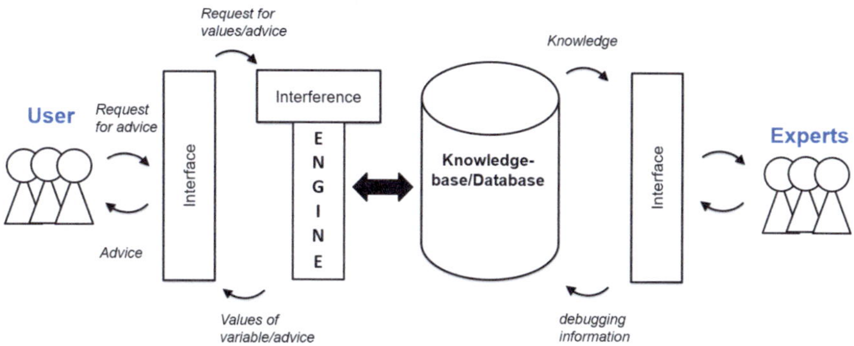

Fig. 3.4 Knowledge-base architecture

is the simplest form of logic. A proposition is a declarative statement that is either true or false in a symbolic variable. For example, P = "you have my car key". Logical connectives are used to connect two symbolic variables logically that operate similarly to arithmetic operators (such as NOT, AND, OR). A powerful Logic property is called *Inference*, which can generate conclusions from evidence and facts to create new logic from old logic. Inference can play an important role in proving a sequence of conclusions that leads to the desired goal. For example, (1) "If you have my car key then you can unlock my car". ($P \longrightarrow Q$), (2) "If you can unlock my car, then you can take my stuff." ($Q \longrightarrow R$) Conclusion: "If you have my car key then you can take my stuff". ($P \longrightarrow R$).

Knowledge-based AI is an agent that emulates how humans think logically. The agent can maintain an internal state of knowledge, reasoning, inference, update their knowledge after observations and take action. For example, the typical KB architecture is shown in Fig. 3.4, experts input their knowledge into KB, and a user asks for advice via a user interface to Inference Engine to deduce from KB. Suppose the environment and the agent can be defined in logical ways. Then, the agent can use knowledge to conclude and solve the problem by forming sentences about the world and using the inference mechanism to infer new sentences to decide what action to perform.

The first method to solve in KB is propositional logic consists of proposition symbols and logical connectives and uses a model-checking inference algorithm, including backtracking and local search methods, to solve the problem. Knowledge of environment and percepts will be converted to propositional logic and inference from KB to the next actions. Example for local search methods can be applied to propositional logic problem spaces; *GSAT* [1], *WALKSAT* [2] and backtracking *DPLL* [3]. However, propositional logic can only represent either true or false or insufficient facts to represent the complex sentence statements, relations, and properties. For example, Alax *likes* a dog. Besides, the model-checking inference is quite slow; with the current commodity hardware, DPLL backtracking algorithm could solve only problems with 30 variables in ten minutes.

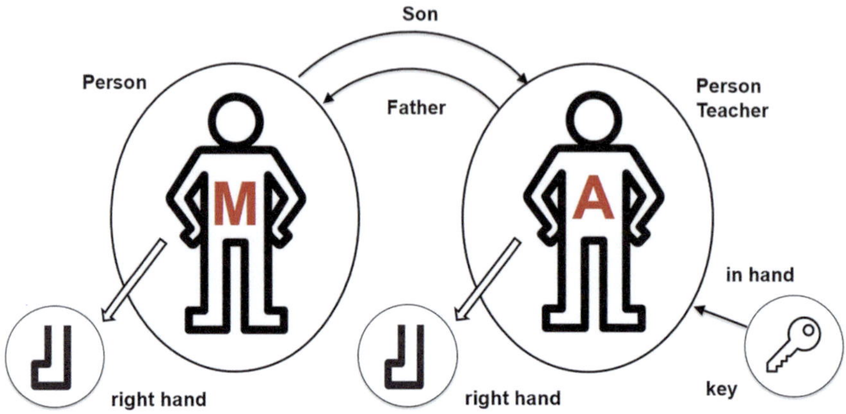

Fig. 3.5 A model that contains objects (Person, Teacher, Key), relations (Father, Son), and properties (right hand)

To handle these issues, First-Order logic (FOL) [4] is introduced, containing relations and their properties. Figure 3.5 shows a model that comprises objects, relations, and properties that can interpret into symbols and grammar. It can map constant symbols to objects, similar to initiating objects from a class in Object-oriented programming (OOP). For example, *Teacher(Alex)* and *Person(Max)* refer to Alex as Teacher and Max is Person. *Father(Alex, Max)* means Alex is the father of Max. The subsequent development of First-Order logic is Knowledge Representation or Knowledge Base (KB), which introduces the idea of a general ontology to organize everything in the world into a hierarchy of categories. A general framework is proposed and called *an upper ontology* that copes in categories, subcategories, parts, structured objects, measurements, substances, events, time and space, change, and beliefs. Even the current framework is robust. However, creating a general framework is not simple and has to be fully realized. KB can be inferred as the first-order logic. Semantic networks [5] also proposed to use KB but provided graphical aids for visualizing a knowledge base and efficient algorithms for inferring the properties of an object. Nevertheless, semantic networks lack logic-based semantics. Description logic (DL) systems [6] evolved from semantic networks to overcome this issue while retaining taxonomic structure as an organizing principle.

Construction of KB is called *knowledge engineering*, which is a process that depends on experts in domain knowledge as Fig. 3.6. There are various working projects in creating KB, for example. CYC system [7] is created by a team of trained ontologists or logicians. DBpedia [8] is built from structure facts in Wikipedia. TEXTRUNNER [9] by parsing from a large corpus or web pages. OPENMIND system [10] is created by volunteers who proposed facts in English. Freebase [11] mainly harvested from various sources, including Fashion Model Directory, Wikipedia, MusicBrainz, and NNDBand.

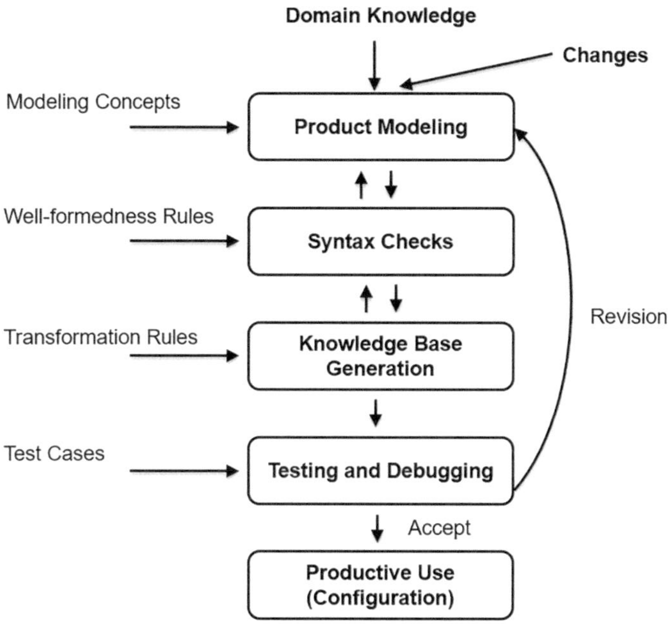

Fig. 3.6 Knowledge engineering process

3.3 Search Algorithms

Search algorithms are algorithms that can be used to solve a wide range of problems or act as rational agents or problem-solving agents. These algorithms are best suited for problems where the complete model is known. Search algorithms can be broadly categorized into two types: uninformed search (blind search) and informed search (heuristic search).

- Uninformed search is a brute-force method that does not utilize information about the state or search space, and only relies on traversing the tree to identify leaf and goal nodes. Because of this, it is also referred to as *blind search*. This method searches through the search space for all possible solutions without any additional knowledge.
- Informed search is a type of search algorithm that uses domain-specific knowledge to guide the search process. This knowledge can include information about factors such as the distance from the goal, path cost, and how to reach the goal node. By leveraging this information, agents can explore the search space more efficiently and locate the goal node more quickly. Informed search employs a heuristics function that takes the current state as input and calculates how close the agent is to the goal node. While this approach may not always guarantee the best possible solution, it can often find a suitable solution within a reasonable amount of time.

The environment is episodic, single-agent, fully observable, deterministic, static, discrete, and completely known. A search algorithm can be used to determine what the agent can perform. In this case, the agent can estimate its actions from the start to its goal with no percepts; for example, in the vehicle routing problem as Fig. 3.2 that the agent needs to know the shortest path to go along. If it can access only the problem definition and no the heuristic function, it uses uninformed search methods, such as Best-First Search [12], Uniform-Cost Search [13]. Nevertheless, If it can access a heuristic function that can estimate the cost of a solution. The map and each place are provided with a heuristic function. Then, it uses informed search methods. For example, A* Search [14]. Each method's difference impacts the solutions' time and quality.

In deterministic environments, an agent can determine its state after taking an action. However, in non-deterministic environments, the agent cannot determine the state it will transition to after taking an action, and thus it requires plans for multiple possible states or belief states. And-Or Search [13] can be used to generate contingent plans in such situations. When the agent's percepts provide only partial observation, a solution can be expressed as a conditional if-then-else expression instead of a sequence of actions. The agent will then test the condition and execute the appropriate branch of the conditional. The agent also maintains its belief state to compute the new belief state from the previous one, rather than relying on the entire set of percepts. Belief-state AND-OR search [15] can be used to solve problems in such scenarios.

In a multiagent environment, there are more than one agent, and each agent can be an adversary of the other agents and compete against them. Each agent needs to consider the actions of other agents and the effect of those actions on their performance. In such scenarios, search algorithms that involve two or more players exploring the same search space with conflicting goals for the solution are called *adversarial searches*. Examples of such algorithms include the Mini-Max Algorithm [16] and its improved version, Alpha-Beta Pruning [17], which enhance their performance speed.

3.4 Machine Learning

Machine learning is a sub-field of AI that studies the ability to improve performance based on experiences. Sometimes, there are only data existing or we cannot provide all possible future situations, we do not know precisely how the program should work, or the task cannot be modeled. Machine learning is an approach that uses data to learn, builds a model, and improves agent behavior or solves problems based on that model. Machine learning can be categorized into three types.

- **Supervised learning** learns from the training data set containing inputs and outputs, labeled as a teacher supervising the learning process. If the output is a category, called a classification problem. However, if the output is a real value, called a regression problem.

- **Unsupervised Learning** contains only inputs but no output variables. It uses to discover and present an interesting structure in the data.
- **Reinforcement Learning** learns by trial and error to maximize an accumulative of rewards.

When humans classify information into categories, we typically begin by identifying the most important feature that best separates items into distinct groups. We then repeat this process, continuing to split based on the next most significant feature, until the remaining groups are small enough to be clearly associated with a specific category.

This is exactly how a *decision tree* [18] works. A decision tree is a model shaped like a tree, where data is split step by step according to selected attributes. Each decision about where to split is guided by a measure called *Information Gain*, which helps identify the most effective attribute for dividing the data at each level.

Information Gain relies on the concept of *entropy*, which comes from information theory. Entropy measures the uncertainty, impurity, or mixing of class labels within a dataset. When all items in a group belong to the same class, entropy is zero, reflecting complete certainty and no label mixing. In contrast, if class labels are evenly distributed across different classes, entropy is high, indicating greater uncertainty or label mixing. The goal of decision tree algorithms is to find an attribute that, when used to split the data, creates child nodes with low entropy or low label mixing. Because each attribute can produce multiple child nodes, we calculate the weighted average entropy of all child nodes after the split.

To identify the best attribute for splitting, we first calculate the entropy of the entire dataset, referred to as the *parent node* entropy. Then, for each candidate attribute, we simulate the split and compute the entropy of each resulting group, called the *child nodes*. These entropies are combined using their weighted average, based on the size of each group. Finally, the *Information Gain* is determined by subtracting this weighted average from the parent entropy. The attribute with the highest Information Gain is selected for the split because it achieves the greatest reduction in uncertainty or label mixing. This process is repeated recursively until the data is divided into subsets that are sufficiently pure to be assigned a class label with confidence. Figure 3.7 illustrates this concept.

$$\text{Information Gain} = \text{Entropy(Parent Node)} - \text{Average Entropy(Children)} \quad (3.1)$$

Although decision trees are simple, scalable, and support both discrete and continuous data, they also have limitations. They use a greedy search strategy, making decisions step by step without considering the broader structure, which may lead to suboptimal accuracy. They are also unstable; a small change in the input data can lead to a completely different tree. Furthermore, deep decision trees can be slow to evaluate and are prone to *overfitting*, especially when the sample size in

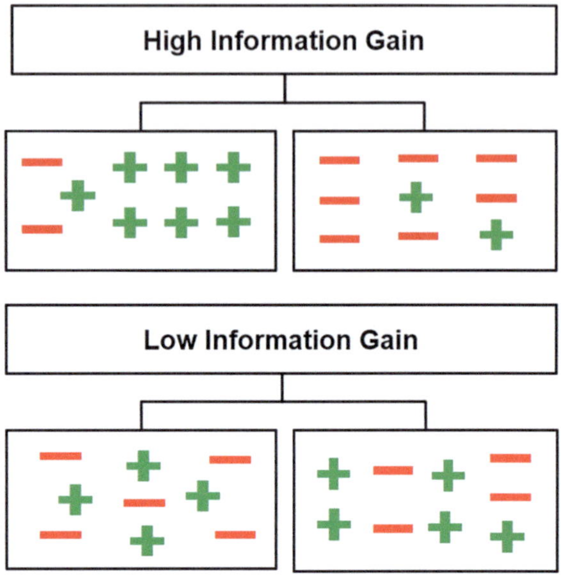

Fig. 3.7 The top picture shows high information gain that can separate the positives and the negatives. Nevertheless, in the below picture, the signs distribute on both sides

some branches is small. While setting a maximum depth can reduce overfitting, this approach introduces bias and often sacrifices predictive performance.

To address these challenges, researchers developed the idea of combining multiple models to form a stronger and more accurate predictive model. This approach is called an *ensemble method*. Ensemble models are grouped into two main categories based on their training approach: *parallel ensembles* and *sequential ensembles*. In parallel ensembles, each model is trained independently of the others. This allows for efficient training, as all models can be trained at the same time. An example of this is a *homogeneous ensemble*, where all base models are of the same type.

In 1994, Leo Breiman introduced *bagging*, short for *bootstrap aggregating* [19]. This technique improves the performance of machine learning models by training multiple versions of the same model on different training datasets generated through *sampling with replacement* from the original dataset. Each of these new datasets, known as *bootstrap samples*, contains the same number of examples as the original dataset. However, because sampling is done with replacement, some examples may appear multiple times in a sample, while others may not be included at all. Each model is trained on a different sample, introducing diversity in their learning. After training, their predictions are combined, typically by majority voting for classification or averaging for regression. Bagging helps reduce variance and prevents overfitting by averaging out the individual errors of the models, making the overall prediction more stable and reliable.

Building on this idea, Breiman later proposed the *Random Forest* algorithm [20], which creates multiple decision trees. To ensure even more diversity among the trees, a random subset of features is selected at each split, instead of evaluating all features.

This additional randomness forces different trees to consider different attributes, which improves the overall diversity and performance of the forest. Random Forests were considered among the best shallow learning algorithms up to around 2014 and are still widely used in practice.

Another type of parallel ensemble is a *heterogeneous ensemble*, where base models are trained using different learning algorithms. The predictions from these models are combined in two common ways. The first is through *weighting methods*, where each model's prediction is assigned a weight according to its performance. Better models receive higher weights and thus have a stronger influence on the final prediction. The second approach uses *meta-learning*, where a new learning model is trained to combine the outputs of the base models in order to make the final prediction.

The second main type of ensemble method is the *sequential ensemble*. In this approach, models are trained one after another, and each new model aims to correct the errors made by the previous ones. A well-known example is *boosting*, which combines many weak learners into a stronger one. One of the earliest and most popular boosting techniques is *AdaBoost* (Adaptive Boosting) [21], introduced by Freund and Schapire in 1995. While parallel ensembles often use strong learners, boosting focuses on *weak learners*, which are simple models that perform only slightly better than random guessing.

In AdaBoost, each training example is given a weight, and examples that are misclassified receive higher weights. The next model focuses more on these difficult examples. This process continues, with each model trying to improve upon the errors of the last. AdaBoost minimizes an *exponential loss function*, which limits its flexibility and makes it sensitive to noisy data and outliers. When misclassified examples are repeatedly emphasized, it may lead to overfitting.

To address these issues, Jerome Friedman proposed *Gradient Boosting Machines* (GBM) in 2001 [22]. Gradient Boosting treats boosting as an optimization problem. Instead of reweighting samples heuristically like AdaBoost, Gradient Boosting directly minimizes a chosen loss function using *gradient descent*. Each new model is trained to predict the *residuals*, or the differences between the true values and the current model's predictions. Unlike AdaBoost, which does not explicitly optimize a loss function, Gradient Boosting systematically improves the model using optimization techniques. Since the mid-2010s, Gradient Boosting, particularly its efficient implementation *XGBoost*, has become the most accurate and widely used method for structured or tabular data. It has consistently outperformed other models in data science competitions and real-world applications. Ensemble methods such as *bagging*, *Random Forests*, and *boosting* offer powerful solutions by combining multiple models to produce more accurate and stable predictions. These techniques, including bagging and boosting, are collectively referred to as *ensemble technique*, as depicted in Fig. 3.8.

Linear functions [23] also be used for classification that creates a linear separator line that separates two classes. The linear function uses training data to estimate into a set of variables (w), called *parametric model*. After learning from the data, the training data can be removed. With the parametric model, we have assumptions to simplify the function to a known form, making the learning process easier. It helps

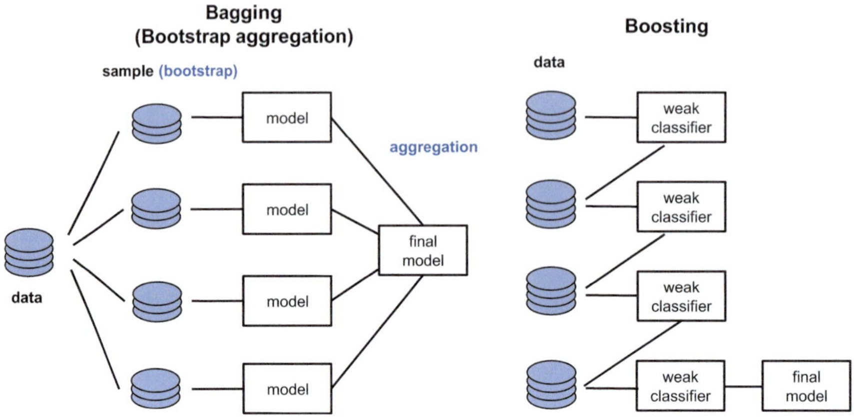

Fig. 3.8 Begging versus boosting technique

in speed, uses less data, and is easier to understand and interpret. However, it can limit what can be learned.

Another approach calls *non-parametric model* that does not make assumptions about the function to a known form. It does not need prior knowledge, no worry about choosing the right features. The simplest non-parametric learning method is table lookup, which keeps all x,y into a table when asked for $h(x)$ and returns y. However, if there is no x in the table, it cannot return y. Nevertheless, instead of finding exactly x, we can find a sample that is the nearest vector to x, and to be more robust, we can find k samples that are the nearest vectors to x, called *k-nearest neighbors* (kNN) [24]. The assumption is that similar points share similar labels. For classification, if $k = 3$ and the three nearest vectors of an input vector represent output classes are (A, B, A). If the number of class A is more than B, then the classification output is class A. kNN works well in low-dimensional space. Nevertheless, in high dimensional spaces, the data space's size grows exponentially with the number of dimensions. In other words, the data must grow to keep the same density. If samples are scarce, the distance among samples is large, and data cannot be separated into classes. In kNN, measuring all the distances and data also takes a lot of time and memory.

Support Vector Machine (SVM) is a nonparametric learning approach that constructs a decision boundary, specifically a linear separating hyperplane, similar to linear or logistic regression. However, instead of relying on a collection of parameter values (w), SVM uses only a subset of training examples, called *support vectors*, rather than all sample points. SVMs aim to find the optimal hyperplane that separates data points from different classes by maximizing the margin between the classes while minimizing classification errors.

In two-dimensional space, this hyperplane is a line that divides the data points into two classes. In N-dimensional space, it becomes an (N–1) dimensional hyperplane. As illustrated in Fig. 3.9, the optimal linear separator maximizes the margin between classes and is defined by the three support vectors closest to the boundary. The

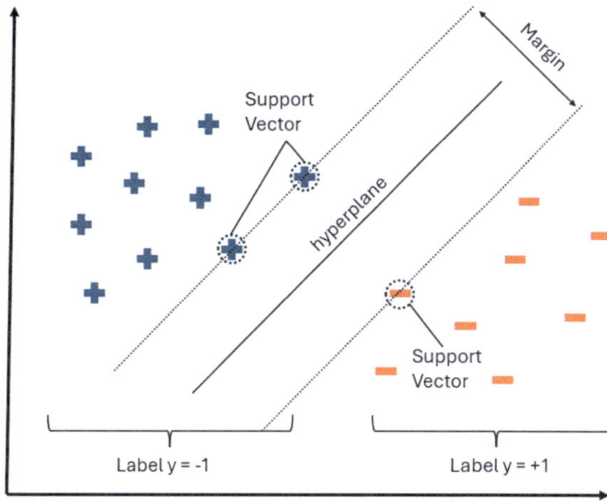

Fig. 3.9 Candidate linear separators and the maximum margin separator (red line) including three support vectors are the examples closest to the separator

key insight is that some data points are more critical than others in determining the decision boundary, which helps improve generalization. When the data is linearly separable, a straight line (or hyperplane) can effectively separate the classes, and in such cases, a *Linear SVM* can be used.

Support Vector Machines (SVMs) aim to find the optimal hyperplane that separates data points from different classes with the maximum possible margin. This hyperplane is defined by a weight vector $\mathbf{w}$ and a bias term b, resulting in the decision boundary:

$$\mathbf{w} \cdot \mathbf{x} + b = 0 \tag{3.2}$$

The margin boundaries, given by:

$$\mathbf{w} \cdot \mathbf{x} + b = \pm 1 \tag{3.3}$$

Create a separation zone between the classes. The width of this margin is:

$$\frac{2}{\|\mathbf{w}\|} \tag{3.4}$$

Maximizing the margin is mathematically equivalent to minimizing the norm of the weight vector. For computational convenience, the objective is formulated as minimizing the squared norm:

$$\frac{1}{2}\|\mathbf{w}\|^2 \tag{3.5}$$

This leads to a convex optimization problem, which is not only easier to solve but also guarantees a unique and globally optimal solution.

The final optimization formulation for the hard-margin linear SVM is:

$$\min_{\mathbf{w},b} \frac{1}{2}\|\mathbf{w}\|^2 \quad \text{subject to} \quad y_i(\mathbf{w} \cdot \mathbf{x}_i + b) \geq 1 \quad \text{for all } i. \tag{3.6}$$

For linear SVMs, after the primal form optimization using algorithms such as coordinate descent or stochastic gradient descent, the output consists primarily of the weight vector (w) and the bias term (b). These parameters together define the decision boundary, which can be used to classify new data points by determining on which side of the hyperplane they lie. Additionally, the algorithm identifies the support vectors, data points that are closest to the hyperplane.

However, in real-world scenarios, data is rarely perfectly separable. To handle such cases, SVMs introduce the concept of a *soft margin*, which allows certain misclassifications while still aiming to maximize the margin. This is achieved by introducing slack variables $\xi_i \geq 0$ for each training instance to permit margin violations. The optimization objective is then modified to balance the trade-off between maximizing the margin and minimizing classification errors. This trade-off is controlled by a regularization parameter C. The modified optimization problem becomes:

$$\min_{\mathbf{w},b,\xi} \frac{1}{2}\|\mathbf{w}\|^2 + C\sum_{i=1}^{n} \xi_i \tag{3.7}$$

$$\text{subject to} \quad y_i(\mathbf{w} \cdot \mathbf{x}_i + b) \geq 1 - \xi_i, \quad \xi_i \geq 0 \tag{3.8}$$

The parameter C controls how strongly the model tries to avoid misclassifying the training data. If C is set to a large value, the model heavily penalizes misclassifications, creating a tighter decision boundary with fewer mistakes on the training set, but this can cause overfitting. In contrast, if C is set to a smaller value, the model tolerates more misclassifications and focuses on creating a wider margin, which may help it generalize better to new data. For example, setting $C = 1000$ will make the decision boundary fit closely to the data, while setting $C = 0.1$ will create a smoother boundary that can handle noisy data more effectively.

However, when the data are non-linear and cannot be separated by a straight line, a *Non-Linear SVM* projects the data into a higher-dimensional space to increase the chance of linear separability, as illustrated in Fig. 3.10. This approach is supported by Cover's Theorem (1965), which states that a complex classification problem is more likely to become linearly separable after a nonlinear transformation into a higher-dimensional space. However, this is a probabilistic advantage, not a guarantee.

To achieve this transformation efficiently, SVM employs kernel functions such as the Radial Basis Function (RBF) kernel, which implicitly maps data into an *infinite-dimensional* feature space. In theory, for any finite set of distinct training points, the RBF kernel can perfectly separate the data by selecting a sufficiently small bandwidth parameter γ. This makes the RBF a *universal kernel*, capable of representing highly complex decision boundaries. Nevertheless, perfect separation of training data does

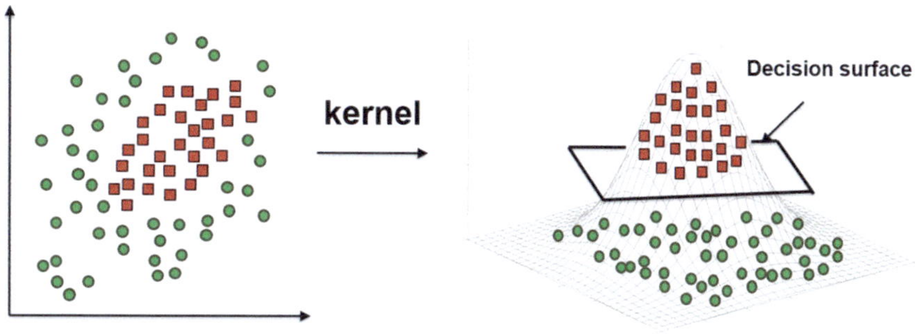

Fig. 3.10 Kernel trick is a simple method that converts from the low dimensional space into a higher dimensional space

not guarantee good generalization to new, unseen data. Therefore, appropriate tuning of γ and the regularization parameter C is essential to control model complexity and prevent overfitting.

To project data into a higher-dimensional space, we can directly create new features based on the existing ones. This transformation from a lower- to a higher-dimensional space can be achieved using techniques such as polynomial expansion or kernel functions (e.g., RBF), which effectively add new dimensions to each data point. In the case of a polynomial expansion of degree d applied to n input features, the number of resulting features is given by the binomial coefficient $\binom{n+d}{d}$. For example, applying a polynomial of degree 3 to 100 input features results in $\binom{100+3}{3} = 176{,}851$ new features. However, Explicitly transforming data into a higher-dimensional space can be computationally expensive and impractical, especially when the number of features grows rapidly. To overcome this, the *kernel trick* is introduced, a powerful technique that allows SVMs to operate in high-dimensional spaces without directly computing the transformed coordinates. This is particularly useful for handling non-linear classification problems where a linear separator does not exist in the original feature space.

The kernel trick becomes applicable during the optimization phase of the SVM. As shown in Eq. 3.6, the SVM training process involves a constrained optimization problem. However, solving the objective function directly is challenging due to the presence of these constraints. To address this, we introduce *Lagrange multipliers*, which are a mathematical tool used to convert the constrained *primal problem* into an equivalent but more tractable *dual problem*.

The advantage of this dual formulation is twofold: it simplifies the optimization and enables the integration of kernel functions. By expressing the objective function entirely in terms of inner products between data points, we can replace these inner products with kernel functions, which implicitly compute dot products in a higher-dimensional space. This allows the SVM to learn complex, non-linear decision boundaries without explicitly performing the transformation.

The resulting dual form of the hard-margin SVM is:

$$\max_{\alpha} \sum_{i=1}^{N} \alpha_i - \frac{1}{2} \sum_{i=1}^{N} \sum_{j=1}^{N} \alpha_i \alpha_j y_i y_j (\mathbf{x}_i \cdot \mathbf{x}_j) \tag{3.9}$$

To handle non-linear data, we map the original inputs to a higher-dimensional feature space:

$$\mathbf{x}_i \cdot \mathbf{x}_j \quad \xrightarrow{\text{map to higher dimensions}} \quad \phi(\mathbf{x}_i) \cdot \phi(\mathbf{x}_j) \quad = \quad K(\mathbf{x}_i, \mathbf{x}_j) \tag{3.10}$$

Substituting this into the dual form gives:

$$\max_{\alpha} \sum_{i=1}^{N} \alpha_i - \frac{1}{2} \sum_{i=1}^{N} \sum_{j=1}^{N} \alpha_i \alpha_j y_i y_j K(\mathbf{x}_i, \mathbf{x}_j) \tag{3.11}$$

subject to:

$$\sum_{i=1}^{N} \alpha_i y_i = 0, \quad \alpha_i \geq 0 \quad \text{for all } i \tag{3.12}$$

Common kernel functions used in SVMs include:

- **Linear kernel**: $K(\mathbf{x}_i, \mathbf{x}_j) = \mathbf{x}_i \cdot \mathbf{x}_j$
- **Polynomial kernel**: $K(\mathbf{x}_i, \mathbf{x}_j) = (\mathbf{x}_i \cdot \mathbf{x}_j + c)^d$, where c is a constant and d is the degree of the polynomial.
- **Radial Basis Function (RBF)** or Gaussian kernel:

$$K(\mathbf{x}_i, \mathbf{x}_j) = \exp\left(-\frac{\|\mathbf{x}_i - \mathbf{x}_j\|^2}{2\sigma^2}\right)$$

where σ controls the width of the Gaussian and determines the influence of individual training examples.

These kernel functions allow SVMs to model nonlinear decision boundaries effectively by implicitly mapping data into higher-dimensional spaces. This is achieved without explicitly computing the transformed feature vectors, making the training process both computationally efficient and scalable to complex datasets.

Once the dual optimization problem is solved, we obtain the optimal values of the Lagrange multipliers α_i. In practice, most of these values are zero, and only a subset of the training examples, known as *support vectors*, will have non-zero coefficients. These support vectors are the most informative points in the dataset, as they lie closest to the decision boundary and directly influence its position. Solving the dual form typically involves quadratic programming, which can be computationally intensive for large datasets. To address this, efficient algorithms such as *Sequential*

Minimal Optimization (SMO) are widely used. SMO decomposes the large optimization problem into smaller sub-problems that can be solved analytically, significantly improving scalability and training efficiency.

SVMs offer several advantages. They perform well in high-dimensional spaces, especially when the number of features exceeds the number of samples. By maximizing the margin, SVMs often generalize well. The use of kernel functions enables them to handle non-linear problems effectively without explicit feature transformation. Additionally, prediction depends only on support vectors, resulting in compact models. However, SVMs also have limitations. Their performance is sensitive to kernel choice and hyperparameters like C and γ, which require careful tuning. Training can be slow for large datasets, especially with many support vectors. SVMs also lack inherent probabilistic output and are less interpretable than simpler models. Though SVMs were dominant in the early 2000s, they were later surpassed by deep learning methods, discussed in Sect. 3.5.

3.5 Deep Learning

Deep learning is a modeling approach that attempts to replicate networks of neurons in the brain using computational units, which are commonly referred to as *neural networks*. The origins of neural networks date back to 1943 when Warren McCulloch and Walter Pitts proposed a neuron model in which each neuron is either "on" or "off" based on activation from neighboring neurons. In 1949, Donald Hebb developed a connection updating rule to modify the strengths between neurons, known as *Hebbian learning*. In 1960, Henry J. Kelley developed a continuous backpropagation model, which was later simplified by Stuart Dreyfus in 1962. However, it wasn't until 1985 that backpropagation became more efficient.

The first version of the deep learning algorithm came from Alexey Grigoryevich Ivakhnenko and Valentin G Lapa in 1965 by modeling with polynomial activation functions that were statistically analyzed. However, the process of forwarding the best statistically chosen features from each layer to the next layer was slow and manual. In 1979, Kunihiko Fukushima introduced convolutional neural networks that used multiple pooling and convolutional layers. In 1989, Yann LeCun et al. applied the first practical backpropagation algorithm to a deep neural network to recognize handwritten ZIP codes on mail. Long-short term memory (LSTM) for recurrent neural networks was developed by Sepp Hochreiter and Juergen Schmidhuber in 1997.

Around 2000, a major issue with deep learning, the vanishing gradient problem, was discovered during backpropagation with gradient-based learning methods. The gradient would become vanishingly small, preventing weight changes in lower layers. This issue can be addressed by using other activation functions such as ReLU.

Even though the fundamental algorithm was developed in 2000, there are many reasons why it took off after 2012. It is because raising of high-performance hardware and large data. Training a model in deep learning requires a large data set, and the

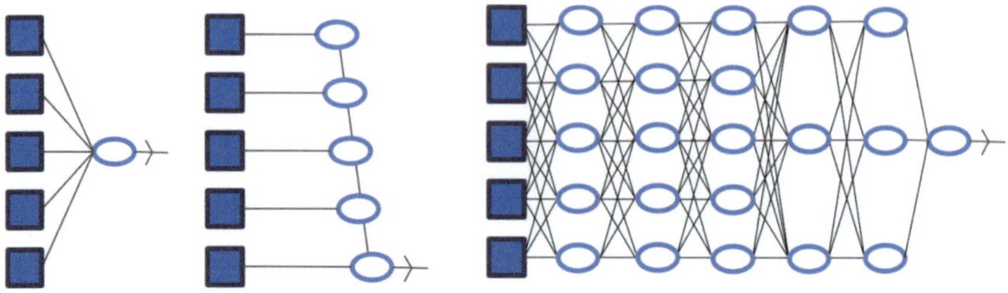

Fig. 3.11 Comparison among linear model, decision tree and deep learning

architecture is complex and requests more computational power than a laptop can do. In 2007, NVIDIA launched CUDA, a programming interface for GPU. Even though GPU is simpler than the CPU, GPU has many simple cores that allow parallel computing that is suitable for matrix operations. GPU is bandwidth optimized to fetch large amounts of memory, such as matrix multiplication. Raising the internet makes it possible to collect and provide large data sets for deep learning, including user-generated tags on public websites, which is a treasure for computer vision.

A comparison of machine learning and deep learning can be seen in Fig. 3.11; the left picture represents a linear model that can handle multiple inputs, but the computation path is short, multiplication with only a single weight, and inputs contribute to outputs independently. In comparison, the decision tree in the middle picture has a long path but is related only to a small fraction of the inputs. The model that focuses on learning only one or two is called *shallow learning*. The right picture presents a deep learning network with a long path, and each input and variable interact. It helps deep learning handle a complex model over traditional machine learning. Deep learning takes advantages over other approaches as the following.

- The needlessness of the feature extraction—Deep learning does not need the feature extraction step. It can learn an implicit representation of the raw data directly by using optimization from backpropagation. Feature extraction reduces too large inputs by identifying key features without losing important or relevant information. It is the key to effective model construction in machine learning. For example, to detect the face of humans, the old method might use the Histogram of Oriented Gradients (HOG), a feature from image data, and learning by using SVM. Instead, Input to the deep learning is image pixels directly.
- Performance can be improved by increasing amounts of data—As Fig. 3.12, in traditional machine learning, performance grows as a power law and reaches a stable line. While deep learning, there is research that shows performance increasing scales along with data size growing.
- Transfer Learning—it is a technique used in deep learning to transfer layers from pre-trained models to a new task, thereby reducing development time and improving performance. Unlike traditional machine learning, which assumes that the training data's distribution characteristics and input feature spaces are in the same domain, transfer learning allows a high-performance learner to be created for a

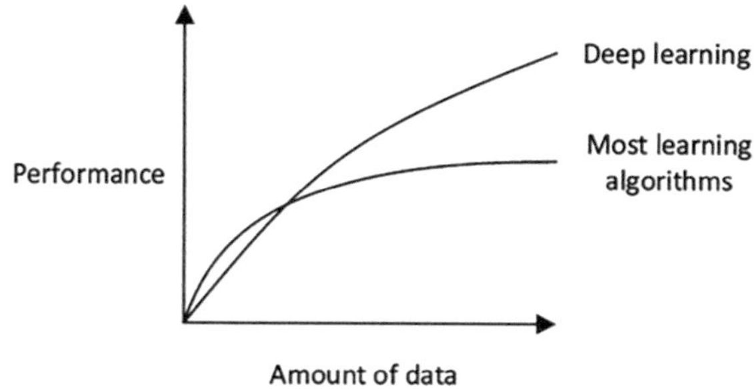

Fig. 3.12 Performance comparison between deep learning and traditional machine learning with the growth of data size

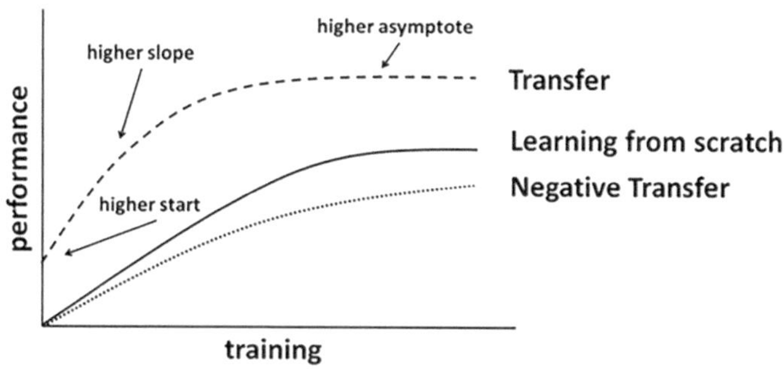

Fig. 3.13 Performance improvement with transfer learning

target domain trained from a related source domain. This approach is motivated by the need to address limitations in target training data, which may be rare, expensive to collect, or inaccessible. Transfer learning can help improve baseline performance, model development time, and the final result. Figure 3.13 shows that applying transfer learning can improve performance from the beginning to the final result.

Machine learning uses input and output to create a model. Similarly, deep learning maps input and output through multiple hidden layers by training many samples of input and targets. Deep refers to layers contributing to a model that can be only two or hundreds of layers. Each layer comprises a basic unit of computation called a node, unit, or neuron. It receives inputs from other nodes in the previous layer or from an external source. A weight (w) is used to assign relative importance from each input to the neuron, sum all weight inputs, and apply the nonlinear function called *activation function*. The activation function introduces non-linearity into the output of a neuron because most real-world data is nonlinear. This mechanism is shown in Fig. 3.14. Thus, deep learning is mapping between inputs and outputs by feeding inputs to each layer; each layer calculates its weights and inputs and then produces the expected output. If we can find the correct weight values for the model,

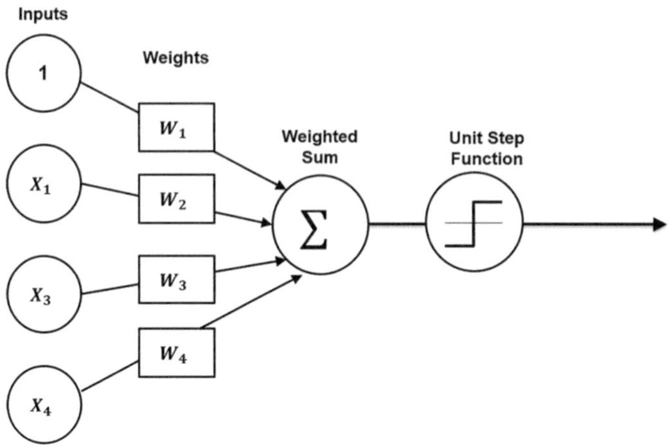

Fig. 3.14 The structure of an artificial neural network

the output will be what we target. However, a deep learning network can contain hundreds of millions of parameters. Finding the correct values are not a simple task, especially since modifying one parameter can impact the whole model. To change weights in the model, we first need to measure how far the current output is from what we expected, called *loss or objective function*. The loss function estimates a distance score, loss score, between the prediction from the current model and the expected value. Deep learning uses this score as feedback to adjust the value of weights from the last layer to the first layer. Weight adjustment is called *optimizer*. The process that uses the feedback to adjust weights is called *backpropagation* shown in Fig. 3.15.

The figure shows the basic architecture of *feed forward neural networks* [25] or Multi-layer perceptron. The feedforward neural network is the first and simplest type of neural network. The information moves from the input nodes, through the hidden nodes, to the output nodes in only one direction with no loops or cycles in the network. Another popular model is *Convolution Neural Networks* (CNNs) [26] that is used mostly in computer vision. The key idea is that it has *filters or kernels* to take some interesting features from input dimensions. The filter is a small matrix that slides over input one by one and for every position. Each filter matrix with different values produces different feature maps for the same input. Finally, the model applies the feed-forward neural network to classify the problem, as shown in Fig. 3.16. Another model of deep learning for sequential data is mentioned in Sect. 3.6.

By 2019, AI systems had reportedly met or exceeded human-level performance in speech recognition, Chess, Atari games, Go, poker, Pac-Man, Jeopardy, translation, Quake III, Dota 2, StarCraft II, skin cancer detection, diabetic retinopathy diagnosis, prostate cancer detection, and protein folding. The term neural network refers to neurobiology or neuroscience, even some concepts in deep learning inspired by the brain. However, deep learning is a model of the brain. It has nothing to do with the brain as it works via mathematical optimization and differential equations.

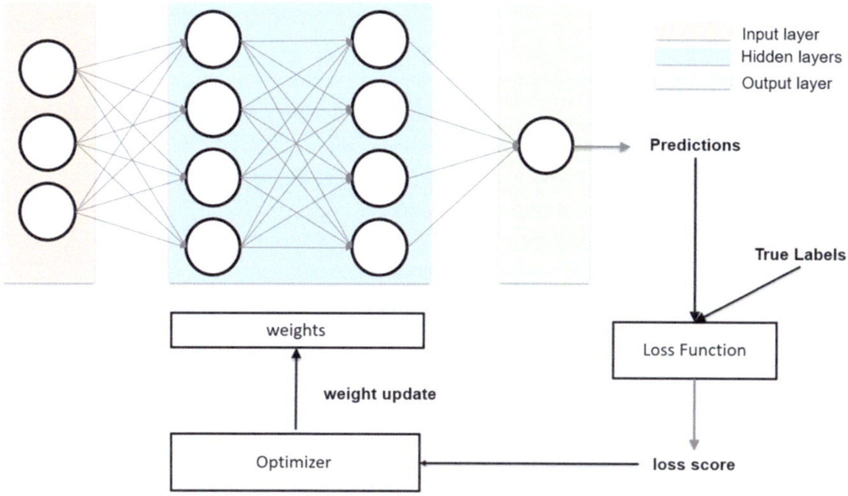

Fig. 3.15 Backpropagation in neural networks

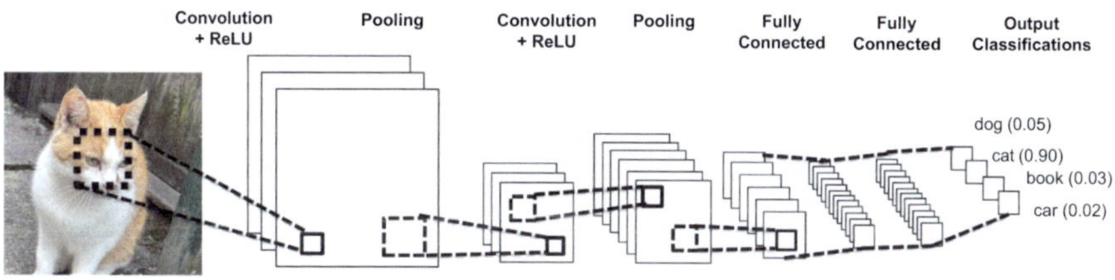

Fig. 3.16 Convolution neural networks

There is no evidence that the brain mechanism works as the learning mechanisms used in modern deep learning models. Deep learning is a mathematical system for learning representations from data. Even though the achievement of deep learning is remarkable but it cannot accomplish in various aspects as the following.

- Optimization and Backpropagation—Deep learning uses backpropagation to modify weights to generate desired outputs. Keeping memory into weights that we do not know exactly how it works creates a black box model.
- Relied on statistics and depended on a lot of data—Even, it can cope with uncertainty. However, it cannot think like a human who can understand and remember or even listen only once.
- Complexity—A complex task needs many layers with different attributes to achieve the task. For example, ResNet, the winner of ILSVRC 2015, can have a very deep network of up to 152 layers with different proposes.
- Request high computation—Deep learning relies heavily on powerful hardware. A standard computer CPU can do 10^9 or 10^{10} operations per second. A deep learning algorithm running on specialized hardware (e.g., GPU, TPU, or FPGA) might

consume between 10^{14} and 10^{17} operations per second, mostly in the form of a highly parallelized matrix and vector operations. It also depends on the availability of large amounts of training data.

- Deep learning models, particularly those that haven't been expressly designed to manage noise, can be extremely susceptible to disturbances in the data. These models inherently lack robustness against noise present in input data. Even minor variations or shifts in the input can trigger substantial changes in the output, which is usually undesirable. Consequently, noise can considerably impair the performance of a deep learning model, leading to erroneous predictions or classifications.
- Do not understand Context—each model is created for a specific task only. Therefore, once changing to different contexts, it will not understand its context.
- Do not understand the meaning—it is very good at mapping inputs to outputs. Nevertheless, it does not really understand the meaning of data and context.

3.6 Sequential Learning

Most problems can be a sequential or temporal environment with relationships between input and time, such as stock, speech, text, or customer behavior. In some problems, we do not know the relation between independent variables, x, and a dependent value, y, or we are unclear about independent variables. One approach is modeling the next step $x(t)$ depending on some previous steps $x(t-k)$. The intention is to understand sequences of information rather than independent or isolated data.

A natural starting point for analyzing temporal patterns is visualization, such as plotting time series to uncover trends, seasonality, or anomalies. However, visual inspection alone may be insufficient when relationships are subtle or obscured by noise. In such cases, statistical decomposition methods can provide a structured understanding. Based on the assumption that a time series comprises trend, seasonal, and random components, decomposition techniques help disentangle these elements to facilitate better forecasting.

There are two primary types of decomposition models: additive and multiplicative. The additive model is appropriate when the seasonal variation remains relatively constant over time, while the multiplicative model is suitable when the seasonal effect scales with the trend. As illustrated in Fig. 3.17, a decomposition of monthly international airline passengers from 1949 to 1960 [27] reveals the distinct trend, seasonal, and residual components. The process typically involves three steps: smoothing with a moving average to extract the trend, detrending to isolate the seasonal pattern, and finally estimating the noise by removing the seasonal component. This preprocessing is often an essential step before advanced modeling, although such assumptions may oversimplify more complex realities.

To forecast future values, one of the simplest approaches is moving average smoothing [28], which computes averages over fixed-length windows of historical

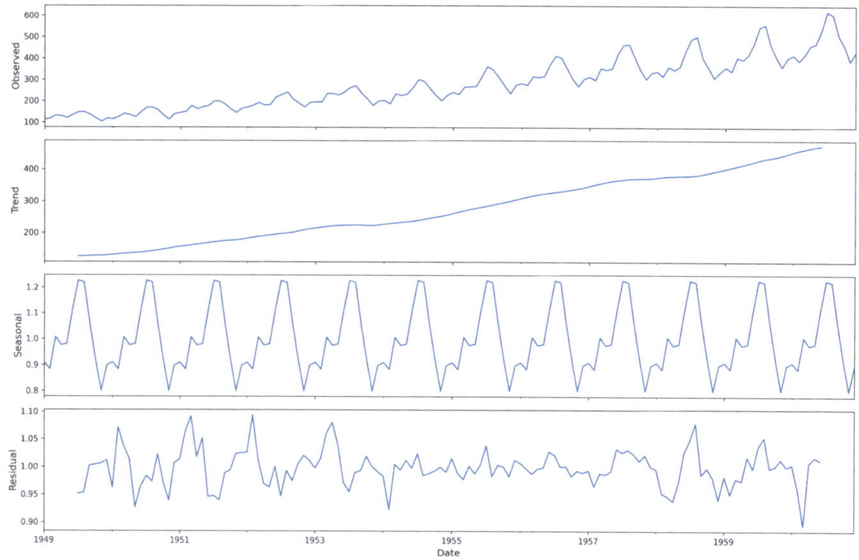

Fig. 3.17 Decomposition of monthly airline passenger data (1949–1960)

data to reduce noise and highlight trends. While effective for short-term smoothing, this method treats all past observations equally. A more refined technique, exponential smoothing, assigns exponentially decreasing weights to older observations, giving more importance to recent data. The assumption here is local stationarity, referring to a slowly changing mean over time. Although effective for short-term predictions, basic exponential smoothing struggles to capture seasonal or cyclical patterns. To address this, extensions such as double and triple exponential smoothing have been introduced.

Despite their simplicity and speed, smoothing methods have limitations in representing autocorrelations and capturing longer-term dependencies in time series data. This is where more robust models such as ARIMA (AutoRegressive Integrated Moving Average) become valuable. ARIMA models utilize the internal structure of the data, specifically patterns in autocorrelation, to produce forecasts.

A key requirement of ARIMA is that the input series be stationary, meaning it maintains a constant mean and variance over time. When the data exhibit trends or seasonal effects, these can be mitigated through a transformation called differencing, as illustrated in Fig. 3.18. A unit root test can be applied to evaluate whether differencing is necessary to achieve stationarity.

Figure 3.18 provides a step-by-step illustration of the differencing process. The first plot displays the original time series, which typically shows upward or downward trends and potentially seasonal variation. The second plot shows the log-transformed series, which helps stabilize variance, especially when the data exhibit exponential growth. The third plot demonstrates seasonal differencing, where values are sub-

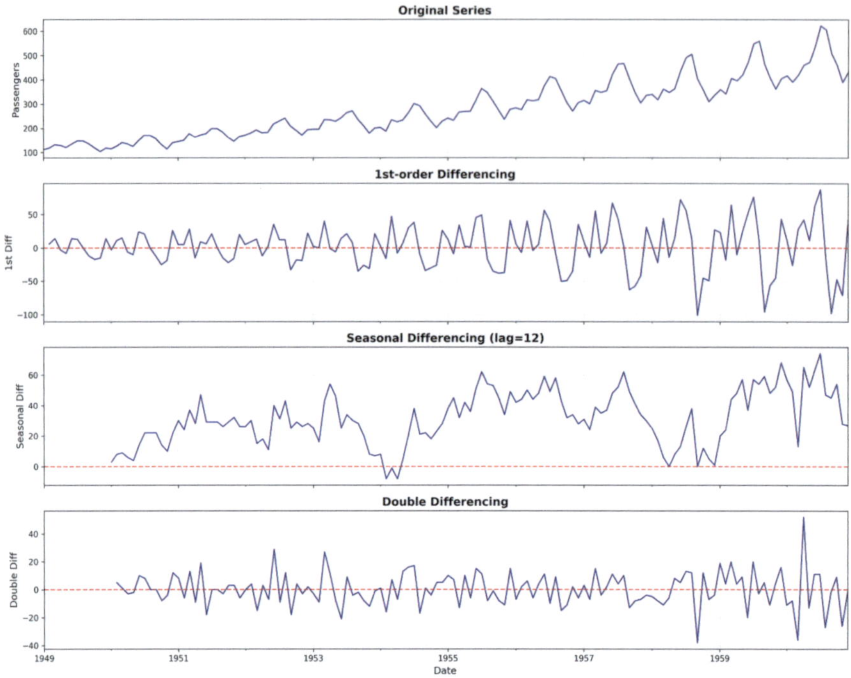

Fig. 3.18 Differencing in ARIMA to achieve stationarity

tracted from those at a previous seasonal lag, effectively removing repeating patterns. Finally, the fourth plot shows the result of double differencing, where both trend and seasonality have been removed. This final series is more likely to meet the stationarity assumption, making it suitable for ARIMA modeling.

ARIMA combines three key components: Autoregression (AR), Moving Average (MA), and Integration (I). The AR component predicts the current value based on previous observations (lags), while the MA component uses past forecast errors. Together, they enable both direct and residual-based forecasting. However, ARIMA assumes linearity, meaning the model captures dependencies as a linear combination of inputs. This assumption limits its applicability in complex or nonlinear scenarios.

To overcome these limitations, neural network-based approaches have emerged, particularly suited for modeling nonlinear and long-range dependencies in sequential data. Models such as the Recurrent Neural Network (RNN) [29] and its advanced variants, including Long Short-Term Memory (LSTM) [30] and Gated Recurrent Unit (GRU) [31], were designed to handle sequences more flexibly. These models maintain internal memory and gating mechanisms that allow them to retain information over longer time spans, making them highly effective in tasks such as speech recognition, machine translation, and time series forecasting.

RNN is the first neural network used to remember the past, and its decision is influenced by what it learned from the past. It differs from the feed-forward neural

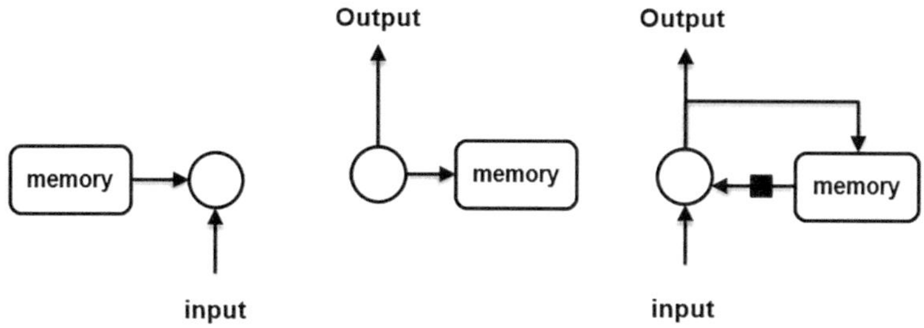

Fig. 3.19 The basic concept of RNN

networks in Sect. 3.5. The output of any layer in the feed-forward does not affect the same layer. While RNN can travel in both directions by looping via the network as feedback. Early inputs are fed back into the network, which helps to keep data in a kind of memory. Unlike feed-forward networks, where weights are fixed once training is complete, RNNs maintain internal states that evolve over time. During inference, while the weights of the RNN remain frozen, the hidden state changes at each time step to reflect the influence of sequential inputs. As shown in Fig. 3.19 in the leftmost diagram, the RNN begins by taking an input and combining it with a stored memory. This mechanism enables the network to incorporate information from previous time steps and process sequential data with context. The second diagram illustrates how this memory is used to generate an output. After producing the output, the internal memory is updated with new information, preparing it for the next time step. The third diagram introduces the concept of temporal recurrence, where a time delay is applied as the updated state is fed back into the system. This feedback loop allows the RNN to capture dependencies across time and connect past and current inputs.

Figure 3.20 presents a more detailed view of the inner workings of an RNN cell. It highlights how each input passes through three main components: the input weight, which scales and transforms the input signal; the memory weight, which determines how the previous state contributes to the current one; and the output weight, which generates the final prediction. These weights are shared across all time steps and are the primary elements adjusted during training.

RNN training uses a technique called backpropagation through time (BPTT). To understand how the network learns, consider the simple sequence "Hello World." During training, the RNN processes this sequence one word at a time. First, "Hello" is fed into the RNN cell. This input passes through the input weight, and since it is the first word, the initial hidden state is typically set to zero. The RNN combines the input with this initial state through the memory weight, generating a new internal state that captures the contextual meaning of "Hello." Based on this state, the RNN predicts the next word by passing it through the output weight. The predicted word is then compared to the actual word "World," and the error is calculated.

This prediction error is used to update the RNN through backpropagation through time. The error signal flows backward through the output, memory, and input weights,

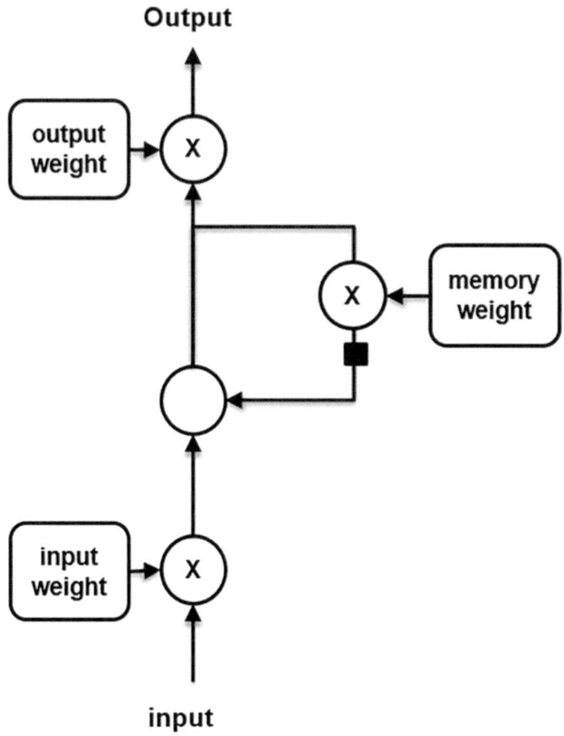

Fig. 3.20 RNN cell architecture

and each of these weights is slightly adjusted to reduce future errors. As a result, the RNN gradually learns the connection between "Hello" and "World."

During inference, the learning phase is complete and the weights are no longer updated. The RNN uses the trained input, memory, and output weights to process new sequences. If the model is given "Hello" as input, it passes through the same flow: the input weight processes the word, the memory weight incorporates the state, and the output weight produces the prediction. This time, instead of learning, the RNN simply outputs the most likely next word based on its training. If trained well, the output will be "World," demonstrating that the RNN has successfully learned the relationship between the two words.

Nevertheless, RNN has problems from BPTT, such as vanishing gradient, exploding gradient, and long-term dependency problems. The first problem is a problem in learning long data sequences. The gradients carry information used to update weights, and when long learning, the gradient becomes smaller and smaller as updating and becomes zero; there's no learning. On the other hand, the derivatives can get bigger and bigger without end; that is the second problem. The last problem is that longer memory is hard to learn through updating the state. The state values depend more on the last state.

LSTM (Long Short-Term Memory) networks were developed to overcome the limitations of traditional recurrent neural networks, especially the vanishing and exploding gradient problems that hinder learning from long sequences. The key innovation in LSTM is the use of gates, which are mechanisms that control the flow of information. These gates allow the model to selectively remember or forget information over time.

This ability enables LSTM to maintain and update memory in a more stable and flexible way, which is essential for tasks involving sequential data such as language processing, time series prediction, or speech recognition. By learning when to retain or discard information, LSTM improves the model's capability to capture long-term dependencies and patterns.

A simplified version of the LSTM, known as the Gated Recurrent Unit (GRU), combines certain gating mechanisms to streamline the model. This design makes training faster and more efficient, particularly on smaller datasets. Both LSTM and GRU greatly improved the performance of sequential learning models by enabling them to better capture complex temporal patterns.

Another major advancement came in 2014 with the introduction of the sequence-to-sequence (Seq2Seq) framework [32], which used an encoder–decoder architecture built on RNNs for machine translation. In this setup, the encoder RNN reads the input sentence and compresses it into a single fixed-length vector. The decoder RNN then generates the output sentence based solely on this vector. This architecture showed that neural networks could successfully map variable-length input sequences to variable-length outputs, opening the door to end-to-end learning in tasks such as translation and summarization.

Despite these advances, RNN-based models including LSTM, GRU, and early Seq2Seq architectures faced several limitations, particularly when dealing with long or complex sequences. A key issue was their difficulty in handling long-range dependencies. Since RNNs process inputs sequentially, information from earlier tokens tends to fade over time, especially in long texts, due to the vanishing gradient problem. While LSTMs and GRUs were designed to mitigate this, they still struggled to retain context over extended input. In addition, the inherently sequential nature of RNNs prevents parallel processing, which leads to slow training and inference, making them less practical for large-scale or real-time applications.

Early Seq2Seq models also introduced a significant information bottleneck. These models compressed the entire input into a single fixed-size context vector, which often failed to capture all the necessary details, particularly in longer sentences. Furthermore, they lacked a mechanism to align specific input parts with corresponding output tokens. In tasks like translation, where each word in the output often depends on specific words or phrases from the input, this absence of alignment made it difficult for the model to focus on the most relevant information during generation. As a result, the model could miss key details or emphasize the wrong content, leading to lower-quality outputs.

To overcome these challenges, researchers developed the Transformer architecture. Unlike RNNs, Transformers do not rely on sequential processing. Instead, they use attention mechanisms that allow the model to consider all positions in the input simultaneously. This enables more effective handling of both short- and long-range dependencies, while also supporting parallel computation. As a result, the Transformer provides a more scalable and context-aware model architecture, which will be explored further in the next Sect. 3.7.

3.7 Transformer and Large Language Models

The Transformer architecture, introduced in 2017 by Vaswani et al. [33], fundamentally reshaped the design of neural sequence models. Diverging from prior sequence-to-sequence (Seq2Seq) models that relied on recurrent layers like RNNs, LSTMs, and GRUs, the Transformer eliminated them entirely. Its power stems from two core components: self-attention mechanisms and feed-forward neural networks. The concept of attention, first introduced in 2015 for neural machine translation [34], laid the groundwork by allowing decoders to dynamically focus on relevant parts of the input sequence, overcoming the limitations of fixed-length context vectors and enabling more accurate predictions for longer and more complex sentences.

Building on this, self-attention was developed, allowing each word within a single sequence to attend to every other word in that same sequence. This innovation enabled models to construct global contextual representations of text and understand relationships between distant words without sequential processing. A key enhancement in the Transformer is multi-head attention. Instead of a single attention mechanism, the model runs several attention heads in parallel. Each head learns to focus on different aspects of the sequence, such as syntactic structure, semantic meaning, or long-range dependencies. The outputs from all these heads are then combined, creating a richer and more comprehensive representation of the input. A significant advantage of the Transformer is its capacity for parallel computation. Unlike RNNs that process tokens one by one, the Transformer computes attention across all tokens simultaneously using efficient matrix operations. This design makes it highly scalable and significantly faster to train, especially with large datasets or long sequences.

The widespread success of the Transformer architecture directly led to the emergence of powerful large language models (LLMs) starting in 2018. Google's BERT [35] utilized a novel pre-training strategy to achieve a deep, bidirectional understanding of context, setting new benchmarks across many NLP tasks. In contrast, OpenAI's GPT-2 [36] focused on text generation using a decoder-only Transformer. With 1.5 billion parameters, GPT-2 demonstrated impressive fluency, coherence, and even zero-shot learning, producing accurate responses to prompts without specific fine-tuning. Subsequent advancements, including models like ELMo [37], ULMFiT [38], and T5 [39], firmly established the importance of large-scale pre-training, contextual understanding, and scalable architectures in modern NLP.

Further development of LLMs saw significant milestones from 2020 onwards. OpenAI's GPT-3 [40] introduced few-shot learning, showcasing that massive models could perform complex tasks like coding, translation, and reasoning from just a few examples in a prompt, without explicit fine-tuning. These were examples of emergent capabilities, skills that only become apparent once a model reaches sufficient scale. GPT-3 underscored that scaling up models unlocks powerful, general-purpose performance. This observation aligned with ongoing research into scaling laws [41], which indicated that increasing model size, data, and computational resources leads to predictable improvements. Larger models also proved to be more sample-efficient, requiring fewer examples to learn effectively. By 2021, various organizations released their own large models (for example, Jurassic-1 [42], Gopher [43], Megatron-Turing [44]), while specialized models like Codex [45] (for code) and LaMDA [46] (for dialogue) also emerged.

Later studies, notably DeepMind's Chinchilla [47] in 2022, refined the understanding of scaling. They found that balancing model size with the amount of training data often leads to better performance, sometimes even allowing a smaller model trained on more data to outperform larger models trained on less. This insight shifted the research focus from merely increasing model size to finding smarter, more efficient training strategies. One such approach was the re-emergence of Mixture-of-Experts (MoE) models in 2021 [48]. These models activate only a subset of specialized sub-networks (experts) for each input rather than the entire model, significantly reducing computational cost. For instance, Google's Switch Transformer achieved 1.6 trillion parameters while maintaining training efficiency and demonstrating 7x speedups compared to dense models. Although more complex to deploy, MoE models offered a scalable path forward without relying on brute-force increases in compute. Concurrently, researchers explored Retrieval-Augmented Generation (RAG) to address limitations like static knowledge and short context windows. Instead of solely relying on memorized data, RAG models retrieve relevant external information during inference. For example, DeepMind's RETRO [49] matched GPT-3's performance using just 7 billion parameters by retrieving helpful context. This approach improved factual accuracy and reduced hallucinations, shifting focus towards equipping LLMs with tools to access external information, a method that gained wider use, supported by frameworks like LangChain [50]. As language models became increasingly powerful, a new critical challenge emerged: ensuring their behavior aligned with human expectations. By 2022, large models could generate fluent and convincing text, but they were still prone to producing inappropriate, biased, or misleading responses. This led to a growing focus on AI alignment, the effort to make models more helpful, honest, and safe (harmless) (the "3H" attributes), and on improving their ability to follow instructions. A major step forward was OpenAI's InstructGPT [51], which combined two key processes: fine-tuning and alignment through Reinforcement Learning from Human Feedback (RLHF). First, the model was fine-tuned using human-written examples to improve its instruction-following capability. Then, using RLHF, the model was further trained to prefer outputs that humans ranked as better responses. This process effectively aligned the model with user intentions, making it more helpful and safe. Remarkably, even a 1.3 billion parameter InstructGPT model outperformed the much larger 175 billion parameter GPT-3 in instruction following. OpenAI launched ChatGPT [52] in late 2022, based on GPT-3.5 and fine-tuned with RLHF and dialogue data. It offered a more conversational, user-friendly experience, quickly reaching 1 million users in five days. Despite its limitations, ChatGPT's success definitively showed that fine-tuning and alignment, not just raw scale, are crucial for making AI practical, safe, and widely adopted.

Now that an overall picture of how LLMs evolved from the foundational Transformer architecture to today's advanced, specialized models has been established, the next section zooms in on the specific architecture of GPT. By detailing each of its core components and mechanisms, it becomes clear how GPT models are structured to process, generate, and understand language in a context-aware, sequential manner.

GPT Architecture

Large Language Models (LLMs) predominantly utilize the GPT (Generative Pretrained Transformer) architecture, which is based on the decoder-only portion of the original Transformer model. This design choice is specifically tailored for autoregressive text generation tasks, where the model predicts the next word in a sequence given the previous context. The training process relies on self-supervised learning, allowing the model to learn to predict the next token without requiring manually labeled data; it leverages the inherent structure of language itself as the training signal, making it both scalable and effective. Modern architectures like Meta's LLaMA [53] and Mistral [54] build upon these foundational GPT principles, introducing incremental improvements while maintaining the core design philosophy that has proven so successful.

To understand the architecture concretely, GPT-3 serves as an excellent reference point due to its impressive scale and specifications:

- It contains 96 Transformer layers, significantly more than the original Transformer's 6 layers, totaling 175 billion parameters.
- The embedding dimension is 12,288, with 96 attention heads processing information in parallel.
- The feed-forward hidden size expands to 49,152, which is four times the embedding dimension.
- The vocabulary encompasses 50,257 tokens, and it has a maximum sequence length of 2,048 tokens.

This massive scale, compared to the original Transformer, reflects the empirical finding that larger models tend to exhibit better language understanding and generation capabilities.

The data flow through the GPT architecture follows a structured sequence of transformations as shown in Fig. 3.21:

Step 1: Tokenization (Inputs → Tokenized): The first step transforms human-readable text into numerical representations that the model can process. GPT models use Byte Pair Encoding (BPE) [55], specifically byte-level BPE, which breaks text into subword units. This approach offers several advantages: it handles unknown words, maintains an efficient vocabulary, and provides cross-lingual capability.

Fig. 3.21 GPT architecture

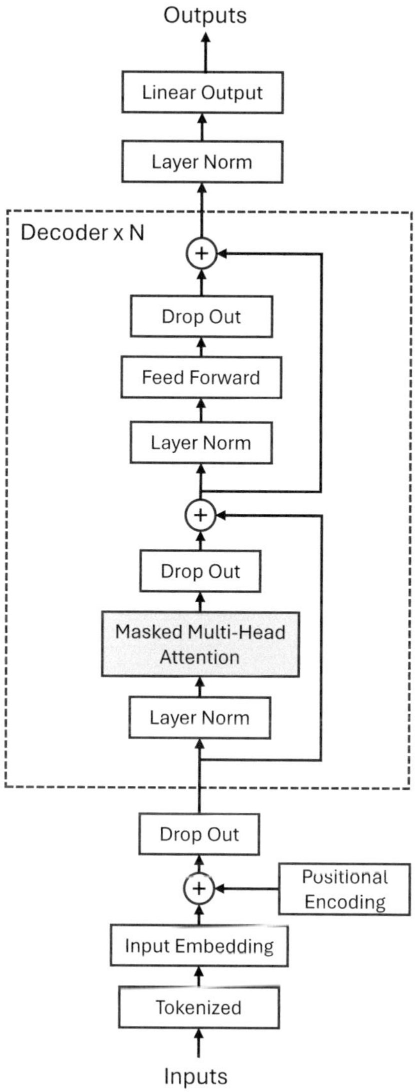

Example tokenization process: Input: "Hello world!" → Tokenized: [15496, 995, 0] Padded to length 10: [15496, 995, 0, 0, 0, 0, 0, 0, 0, 0] Output shape: (1, 10).

Step 2: Input Embedding (Tokenized → Input Embedding): Once tokenized, input embedding converts discrete token IDs into dense vector representations that capture semantic meaning. Each token ID is mapped to a 12,288 dimensional vector through a learned embedding matrix. These embeddings are trainable parameters that evolve during training to encode rich semantic and syntactic information about each token. The transformation takes token IDs with shape [10] and converts them to embeddings with the dimensional shape (10, 12288).

Step 3: Positional Encoding (Input Embedding + Positional Encoding): Transformers process all tokens in parallel and therefore do not naturally know the order of the tokens. Positional encoding adds this information so the model can understand the sequence. Unlike the original Transformer which used fixed sinusoidal encodings, GPT uses learned positional embeddings that allow the model to adapt its positional understanding based on specific patterns in its training data. The positional embeddings are simply added to the token embeddings, with the final input being the sum of token embedding and positional embedding, maintaining the shape of (10, 12288).

Step 4: Dropout (After Positional Encoding): Dropout serves as a regularization technique to prevent overfitting by randomly setting some embedding elements to zero during training. With millions or billions of parameters, neural networks can easily memorize training data, so dropout forces the model to learn robust representations that do not rely on specific neurons, improving generalization to unseen data. Output dimensions remain unchanged as dropout only affects training behavior, not the inference architecture.

Step 5: Layer Normalization (Before Attention): Layer normalization plays a crucial role in stabilizing the training process by normalizing activations to have zero mean and unit variance. This component provides faster convergence by reducing sensitivity to parameter initialization, ensures gradient stability by preventing vanishing or exploding gradients in deep networks, and maintains scale invariance by handling inputs with different magnitudes consistently.

Step 6: Masked Multi-Head Self-Attention: The core of the architecture lies in the masked multi-head self-attention mechanism, which combines four key innovations that make Transformers powerful for language modeling: self-attention, multi-head attention, positional encoding, and masking. The self-attention mechanism allows each token to attend to or gather information from all previous tokens in the sequence. This attention process uses three learned projections: Query (Q) asking "What information am I looking for?", Key (K) representing "What information do I contain?", and Value (V) providing "What information should I provide?"

Mathematical process for each attention head i:

$$
\begin{aligned}
Q_i &= X \times W_Q^i;\ (10, 12288) \times (12288, 128) = (10, 128) \\
K_i &= X \times W_K^i;\ (10, 12288) \times (12288, 128) = (10, 128) \\
V_i &= X \times W_V^i;\ (10, 12288) \times (12288, 128) = (10, 128) \\
\text{Attention}_i &= \text{softmax}\left(\frac{Q_i \times K_i^T}{\sqrt{d_k}}\right) \times V_i
\end{aligned}
$$

Where:

- X: Input embeddings for the tokens in the sequence.
- W_Q^i, W_K^i, W_V^i: Trainable weight matrices for the query, key, and value projections of attention head i.

- Q_i, K_i, V_i: Projected representations of the input used to compute attention.

Once the input X enters the layer, each attention head (i) begins by creating three specific projections: a query (Q_i), a key (K_i), and a value (V_i). These are generated by multiplying the input X with the head's unique, learnable weight matrices (W_Q^i, W_K^i, and W_V^i). The scaled dot-product attention is then computed: this involves taking the dot product of Q_i and the transpose of K_i, scaling the result by the square root of the key dimension (d_k), and applying a softmax function to produce the attention scores. These scores essentially tell us how much each part of the input should "pay attention" to other parts (Fig. 3.22). Finally, these attention scores are used to weight the corresponding values V_i, resulting in the output of that specific attention head, which represents the focused and relevant information extracted from the input.

This process is performed in parallel across all attention heads, and the outputs from each head are then concatenated and passed to the next layer.

The scaling factor $\sqrt{d_k}$ prevents the dot products from becoming too large, which would cause the softmax function to have extremely small gradients that hamper learning. The Query (Q) seeks information (for example, from the word "book"), which then searches for relevant information in each word in the input (Keys, K) by measuring similarity using dot products. When attention scores are available, they are then applied with V, which contains the actual content and features to retrieve. V transforms the input embeddings into value representations that contain the information to be aggregated. Unlike a standard feedforward layer that applies the same transformation everywhere, the attention mechanism selectively weights these values based on the computed attention scores.

During training, gradients flow through each W matrix (for Q, K, V) based on how each contributes to the attention computation and final loss. Through this end-to-end optimization, Q specializes in creating queries that capture what to look for, K specializes in creating representations that are easily matchable with relevant queries, and V specializes in encoding the actual content that will be aggregated. The matrices co-evolve to form an effective attention system.

Multi-head attention provides the model with multiple parallel perspectives. Different heads specialize in capturing different types of relationships: syntactic (for

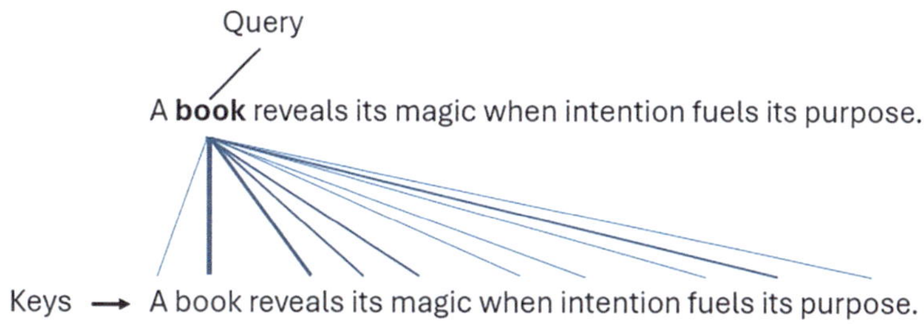

Fig. 3.22 Examples of self-attention: "book" is the query, and attention is computed with all words (keys) in the sentence

example, subject-verb), semantic (for example, synonyms), long-range dependencies, and local patterns. GPT-3 uses 96 attention heads, each operating on a 128 dimensional slice (12,288/96=128), before concatenating all head outputs.

$$\text{multi\_head} = \text{Concat}([\text{head}_1, \ldots, \text{head}_{96}]) \quad ; (10, 96 \times 128) = (10, 12288)$$

Causal masking preserves the autoregressive property critical for language modeling. It ensures the model can only use information from previous tokens, not future ones, by applying a mask matrix that sets future positions to negative infinity (which become zero after softmax). Masking is essential for training consistency, preventing data leakage, and enabling step-by-step autoregressive generation.

Step 7: Dropout (After Attention): After multi-head attention, dropout is applied for additional regularization, specifically to the attention mechanism, to prevent overfitting during training.

Step 8: Residual Connection: A residual connection adds the original input to the attention output. This prevents gradient vanishing by creating gradient highways, allowing information and gradients to flow directly through the network, crucial for training very deep models. This addition maintains original information while incorporating attention-processed features.

Step 9: Layer Normalization (After Residual Connection): Following the residual connection, layer normalization stabilizes the result and prepares it for the feed-forward network, ensuring activations maintain appropriate statistical properties.

Step 10: Feed Forward Network: The feed-forward network applies position-wise processing to each token independently, enabling the model to learn complex feature combinations through nonlinear transformations. It consists of a linear layer expanding from d_{model} (12,288) to $4 \times d_{\text{model}}$ (49,152), followed by a GELU activation function [56], then another linear layer contracting back to d_{model} (12,288). Input and output dimensions remain (10, 12288). This 4x expansion-contraction architecture provides increased representational capacity, and GELU is preferred for its smoother gradients.

Step 11: Dropout (After Feed Forward): Dropout is applied after the feed-forward network to provide additional regularization for the position-wise transformations, helping prevent overfitting in this non-linear feature learning component.

Step 12: Second Residual Connection: Another residual connection adds the input to the feed-forward block to its output, maintaining information flow and gradient stability. This creates a second highway for information to bypass feed-forward processing when needed.

Step 13: Decoder Block Repetition (N Times): The decoder stack consists of N repeated blocks; GPT-3 uses 96 layers. Each block contains the complete sequence from masked multi-head self-attention through the final residual connection. This large number of layers allows the model to capture increasingly abstract and complex patterns, with early layers focusing on simple patterns and local dependencies, middle layers handling syntactic structures and medium-range relationships, and later layers processing complex semantic relationships and world knowledge.

Step 14: Final Layer Normalization: The layer ensures the output of the last decoder block maintains stable probability distributions, smoothing the output distribution to aid final prediction stability and improve overall output quality and convergence.

Step 15: Linear Output Layer (Language Modeling Head): This final linear layer projects the hidden states back to vocabulary space to generate probability distributions over possible next tokens. This translates continuous vector representations back to discrete token probabilities, enabling the model to generate actual words and phrases.

Output generation process: Hidden State: [sequence_length, d_model] = (10, 12288), Linear Projection: (12288) $\rightarrow$ (50257) (vocabulary size), Output: Probability distribution over all possible tokens.

The GPT architecture's robust design, combining scalability, self-attention's flexibility for capturing dependencies, and residual connections to maintain representational power, forms the foundation of most modern LLMs and continues to drive advancements in natural language understanding and generation.

While the architecture provides the structural backbone of models like GPT, their real-world effectiveness hinges on how they're trained and refined. Beyond the initial design, the development of LLMs involves pre-training on massive datasets, fine-tuning for task-specific needs, and alignment tuning to ensure outputs align with human expectations. The following section examines how these critical phases combine to make LLMs truly capable and reliable.

LLM Development

The development of LLMs centers around three core areas: Pre-training, Fine tuning, and Alignment Tuning.

Pre-training focuses on building an LLM from scratch, encompassing choices about datasets, learning objectives, and architectures. GPT-3, for example, used the next-word prediction task, a form of self-supervised learning where the model trains on massive unlabeled text datasets by leveraging the inherent structure of language itself, without needing explicit human labels. GPT-3 was trained on a diverse mix of large datasets: the vast Common Crawl web corpus (66%) [57, 58], OpenAI's internal WebText2 [59] (22%), various book corpora (16%) (OpenAI does not disclose the exact sources, but these are widely based on large English-language book corpora such as BookCorpus [60] and Project Gutenberg [61]), and Wikipedia [62] (3%). The sheer quality and scale of this training data are paramount. As the saying goes, "garbage in, garbage out", and even subtle biases in massive datasets can propagate into systemic issues in model behavior, impacting trust and real-world outcomes.

Even after extensive pre-training, powerful models like GPT-3 often lack precise alignment with user intent. While they can generate coherent language, they might produce factually incorrect, irrelevant, or even offensive content because their training solely optimizes for predicting the next word, not for following instructions or generating safe, helpful responses. This is where fine-tuning and adaptation techniques become pivotal, tailoring foundational models to specific tasks and domains, addressing their general and static nature. Supervised Fine-tuning (SFT) is a common method where a pre-trained LLM is trained on a smaller, high-quality, labeled dataset relevant to a specific task. This allows organizations to adapt cutting-edge models without building them from scratch, significantly reducing computational expenses. However, standard SFT can be computationally expensive as it updates all model parameters and risks catastrophic forgetting, where the model's broad pre-trained knowledge degrades.

To address these challenges, Parameter-Efficient Fine-Tuning (PEFT) methods have gained prominence. PEFT techniques adapt models by updating only a small subset of parameters. LoRA (Low-Rank Adaptation) [63] is a leading PEFT method that adds small, trainable rank-decomposition matrices to the attention and feedforward layers while keeping the original weights frozen. This dramatically reduces trainable parameters (by 1,000 to 10,000 times) and GPU memory usage, often matching or surpassing full fine-tuning performance. Importantly, trained LoRA matrices can be merged back into the main model, adding no inference latency. Another efficiency approach is Compute-Efficient Fine-Tuning through quantization, which reduces memory requirements by representing parameters in lower precision (for example, from FP32 to INT8). Post-Training Quantization (PTQ) [64, 65] is applied after a model has been fully trained. It transforms the model's weights and activations to lower-precision representations without additional training. While PTQ is simple but can reduce accuracy, Quantization-Aware Training (QAT) integrates quantization during training for better results. Building on this, QLoRA (Quantized Low-Rank Adaptation) [66] combines low-rank adaptation with aggressive 4-bit quantization, enabling fine-tuning of large LLMs even on single GPUs. Additionally, Distillation [67] creates smaller student models by transferring knowledge from larger teacher models, offering faster inference with a slight accuracy trade-off.

Finally, alignment tuning addresses the critical issue of ensuring LLM behavior aligns with human expectations. As models become more powerful, they can still produce undesirable outputs (for example, hallucinations, biased language, misinterpretations of instructions). This highlights a gap between the model's default training objective and user desires for helpful, honest, and safe outputs. The goal of alignment tuning is to adjust the model's responses to meet these 3H (helpful, honest, harmless) attributes. Reinforcement Learning from Human Feedback (RLHF) is one of the most successful alignment techniques, notably used in InstructGPT [51] and ChatGPT. In RLHF, human annotators review model outputs and provide feedback (for example, ranking responses). This feedback is used to train a separate reward model, which then guides the LLM (the policy model) via reinforcement learning to produce outputs that maximize the predicted reward, iteratively aligning the model with human values. An alternative, Direct Preference Optimization (DPO) [68], simplifies this by directly optimizing the model based on pairwise human comparisons without needing an intermediate reward model, making it faster and simpler to train. While RLHF excels at handling complex, nuanced feedback for safety-critical uses, DPO offers efficiency for general alignment. Many organizations now adopt hybrid approaches, combining the strengths of both for robust and well-aligned LLM deployments.

In machine learning, these algorithms can map between input and output in an orderly manner that uses training data to generate a model. However, how can the agent learn from trials and errors by an algorithm that knows good or bad behaviors to reward the agent? This approach can be found in the next Sect. 3.8, called *reinforcement learning*.

3.8 Reinforcement Learning

In reinforcement learning (RL), the agent has no rules or data to teach a model to interact with an environment. The agent will try out different actions as a trial and error by the environment. It will give a reward to the agent for every action it does. The agent receives a positive reward when it performs a good action, and a negative reward is provided when it performs a bad action. Thus, It will learn the good action to maximize rewards.

As Fig. 3.23, in step 1, the environment is described in a set of numbers from environment states or states. Then, the agent looks at the environment and takes action. The action is selected from the list using an algorithm called *policy*, along with private information. Then, the agent provides the action to the environment, and the environment will evaluate the action and produce a new state. Sometimes, the agent cannot produce a new state because, in a real environment, it might find unexpected events or the environment is dynamic. For example, the agent acts a at state s is not always go to state $s1$ due to some randomness based on some probability distribution. It is called a stochastic environment. Conversely, when the agent performs action a in state s, and it always reaches state $s1$, then the environment is called *a deterministic environment*. In step 2, the environment updates itself by providing a new state list and uses the new states to determine actions that the agent might choose on its next move, including giving the agent a reward. It also includes environmental information to update for private information and policy. The agent saves its updated private information and policy in the final state. At some point, the agent achieves its final goal and ends its mission; the environment will update the agent with a final reward. This agent interacts with the environment from the initial state until the final state is called *an episode*. The agent learns each episode iteratively, and the policy is improved over time to find the optimal policy that performs the correct action in each state to maximize its reward. Thus, the goal of RL is finding *the optimal policy* or the correct action in each state to maximize the return or sum of rewards.

In RL, the agent takes action to transition from one state to another state that can be represented to *markov reward process* (MRP), which is an extension of *markov decision process* (MDP) with a reward function. The reward function returns the reward that will be received for each state. The markov property defines that the future depends on only the present. It composes of sequences of states and predicts the next state by depending on the current state solely, not the previous state. As

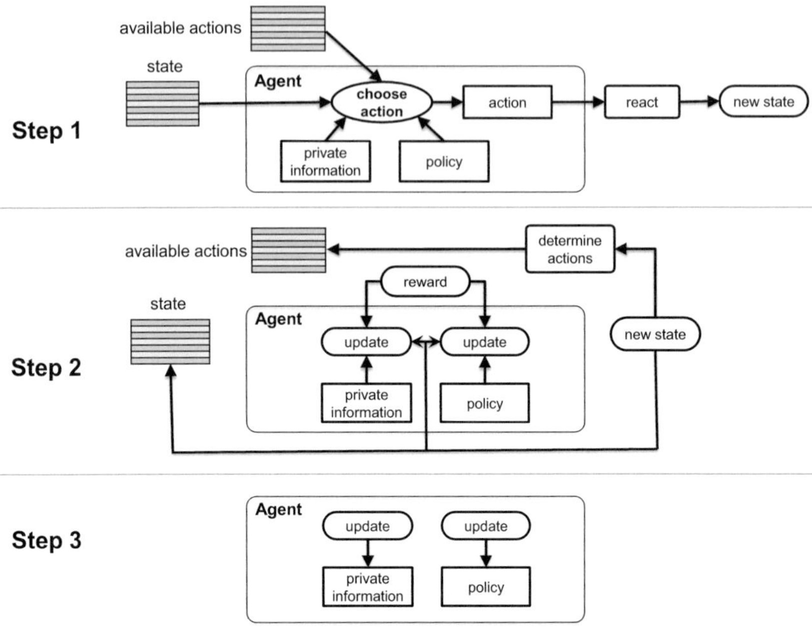

Fig. 3.23 A generalized structure of reinforcement learning

Fig. 3.24, the stable state economy can transit to itself with a probability 0.7, to the recession state with a probability 0.25, and the economy bubble state for 0.05. If it chooses an action to stay in a stable economy, it will be rewarded. *Total reward* is the sum of all the rewards from the first state to the last state of an episode. *Total future reward* is the sum of the current state to the last state. If the current is the fifth state, the total future reward is the sum of the fifth to the last state. Nevertheless, if the task is too long or no a terminal state, the reward is up to infinity or too big. If it has a long episode, it has more variance and includes more irrelevant information. *Discounted future reward* is used to handle this issue as the following equation. It determines how much the agents care about rewards in the future compared to the immediate states by using *discount factor* (γ), as shown in the Eq. 3.13.

$$R(\tau) = \gamma^0 r_0 + \gamma^1 r_1 + \cdots + \gamma^n r_n \tag{3.13}$$

The agent decides to select an action in the current state by using *Q-function* or *state-action value* function. The state-action value function represents the total future reward of each action and state that can be found in the left picture of Fig. 3.25. For example, at the current state $S0$, the agent decides to act $A3$ as it gives the highest reward, 19. Also, *Value function* can be used to evaluate the current situation. The value of each state is the return of the total future reward that the agent would obtain from that current state to the last state by following the policy π that contains actions.

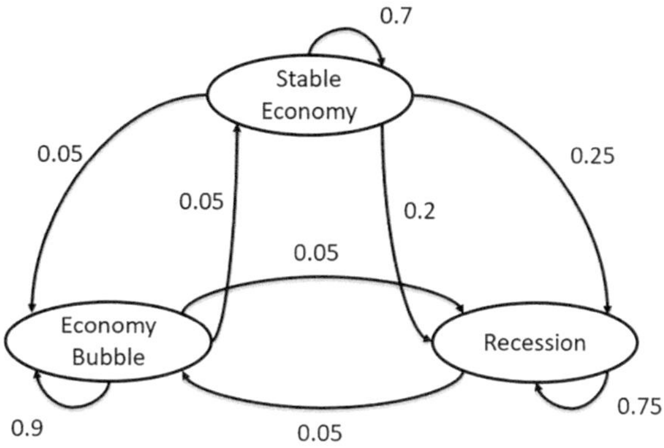

Fig. 3.24 A sample Markov process

Fig. 3.25 The Q-function and the value function examples

State	Action	Value
S0	A1	15
S0	A2	13
S0	A3	19
S1	A1	5
S1	A2	7
S1	A3	9

State	Value
S0	13
S1	5

State	Value
S0	19
S1	9

As Fig. 3.25, the top right picture shows the value function for a policy ($S0 \Rightarrow A2$, $S1 \Rightarrow A1$). The right bottom picture shows another policy ($S0 \Rightarrow A3$, $S1 \Rightarrow A3$) that this policy is also the optimal policy that provides the highest reward.

RL intends to find the optimal policy that provides the best action for each state to maximize the reward. The first method is *dynamic programming approach* (DP) and separate into two methods, *value iteration* and *policy iteration* methods [69]. The value iteration bases on the Eq. 3.14 that is derived from *bellman equation* [70]. $R(s, a, s')$ refers the reward to take action a from state s to state s', $V^*(s')$ is the optimal value function of the next state s'. γ is the discount factor. The environment is stochastic; after the agent takes action a, the state can be transited to any state s' with the transition probability $P(s'|s, a)$. Currently, $P(s'|s, a)$, $R(s, a, s')$ and γ are known. Only the value table, $V^*(s')$, is missing. The method initiates the value table randomly to estimate the Q-function. Once the Q-function is convergence, the optimal policy can be extracted from this Q-function.

$$Q^*(s, a) = \sum_{s'} P(s'|s, a)[R(s, a, s') + \gamma V^*(s')] \tag{3.14}$$

Instead of estimating the Q table from a random value table to find the optimal policy, the policy iteration finds the optimal policy from a random policy. The optimal value function is computed by using the policy iteratively, the optimal value function is found, then the optimal policy can be extracted. This method also is derived from the bellman equation as Eq. 3.15. Currently, $P(s'|s, a)$, $R(s, a, s')$ and γ are known. $V^{\pi}(s')$ and a policy is initiated randomly.

$$V^{\pi}(s) = \sum_{s'} P(s'|s, a)[R(s, a, s') + \gamma V^{\pi}(s')] \tag{3.15}$$

At first, the value function is computed using the random policy repeat until the value function converges. Then, a new policy is extracted from the computed value function. After that, the previous and new policies will be compared; if they are the same, stop. Else, repeat the process to compute a value function again to find a new policy. The advantage of DP is that the value of a state can be estimated based on the value of the following state called *bootstrapping*, with no need to wait till the end of the episode. However, the disadvantage of DP is that the environment must be the model dynamic that knows the transition probability, $P(s'|s, a)$, called *model-based method*.

RL had many successes in various areas. Such as Go, which beat the world champions of the Go board game. RL can solve many complex problems without requesting training data. It can perform actions as a trial and error to find optimal solutions. However, some disadvantages still break RL from using in real life. At first, it requested a lot of time to perform trial and error, but humans only take a few minutes to know. Secondly, RL is not a golden hammer; many other approaches perform even better. Next, RL requires a reward function; sometimes, it is hard to design. The results might be local optima and unstable, which causes weird patterns and is hard to reproduce.

3.9 Criticism of State of the Art

Knowledge Representation (KB) is human knowledge embedded in hard code into the program, graph, or database. It is good in situations with no requirement for data, a danger of errors, and output speed. However, it requested input in a defined structure before manual insertion. Besides, It is almost impossible to manually input all possible logic from real-world problems into the application or database. KB could not adapt itself to changing environment if no rules existed. Humans must insert new logic into the system again.

Search algorithm is finding solutions brute-force or heuristic from problem spaces. This solution is very effective as the problem is fixed. However, it cannot be applied to many cases in real-world problems as the environment is not deterministic.

The third type of AI agent is *Reinforcement learning (RL)* including *Deep Q Network (DQN)* that learns from performing actions as a trial and error and reinforces the correct behaviors by the reward system. It resembles a control system that adapts its parameters as the environment changes. It is good for solving complex problems; no training data set is requested as it learns from experience. However, it requires a lot of data to train by performing trial and error, except most humans can learn in a few minutes. In many cases, some approaches can get better results. For example, using online trajectory optimization [71] to control humanoids in simulation performs even better than RL. Also, the final results of RL can be unstable to weird patterns and hard to reproduce.

The fourth type of AI agent is *Machine learning*. Machine learning uses training data to create a model that maps between input and output. Machine learning easily identifies patterns and trends and improves the result over time. However, machine learning works offline, and the model is trained from training data. Hence, it is missing adaptive in the dynamic environment. *Deep learning* is another machine learning technique that tries to replicate the learning of neurons in the brain. It showed a very good performance. However, it also works in offline mode. It is known as a black block model; primarily due to its use of backpropagation optimization to adjust its weights. The internal workings of this model remain elusive in the research literature. As the Sect. 3.5, deep learning cannot achieve in various aspects, including depending on a lot of data, complexity, and high computation. Moreover, the layer and connections are fixed; this limits learning as the model is created for a specific problem.

Next, *time-dependent learning*, which learns how previous inputs and the current input impact output, comprises ARIMA, RNN, LSTM, and GRU. These algorithms work well in time or sequentially. LSTM and GRU also use long-short term memory that can forget and remember temporarily or promote long-term memory. Even time-dependent learning can adapt itself from previous input, but it is also considered offline learning as it is trained and the model is not changed. Despite their improvements over basic RNNs, these models still face limitations in handling very long sequences due to vanishing gradients and their inherently sequential processing nature, which prevents parallel computation and leads to slower training and inference.

The introduction of the *Transformer architecture* represents a significant advancement in AI capabilities. However, these models still face substantial challenges, including enormous computational requirements that consume energy equivalent to hundreds of homes and demand expensive, specialized hardware. They are also prone to hallucinations, generating plausible but factually incorrect information with high confidence, and lack true understanding or reasoning capabilities, processing text statistically rather than comprehending meaning. Their black-box nature makes it difficult to interpret their decision-making processes or debug problematic behaviors, while their static knowledge remains frozen at training cutoff dates, preventing continuous learning from new interactions. Unlike biological intelligence, which adapts continuously, LLMs operate in an offline learning mode only, unable to modify their behavior based on user feedback or changing contexts during deployment. Additionally, they exhibit sensitivity to input variations, where small prompt changes can lead to dramatically different outputs, and are vulnerable to adversarial attacks, limiting their reliability in real-world scenarios where robustness and consistent performance are critical.

Last but not least, *HTM* proposes online learning and adapts changing to its input. However, it is important to note that HTM is developed by a private company, so it should be cautious when evaluating it. HTM is highly centralized, with significant influence held by Numenta that could result in the company promoting its goals over other considerations. Learning of synapses, dendrites, and segments works in sequence operations can impact its performance on a large scale. Moreover, the applications of NLP in HTM are still limited and not widely used. Additionally, HTM is designed to focus mainly on the neocortex, many other brain attributes are unable to escape integration with the neocortex in order to achieve full intelligence.

In summary, a comparison between techniques and learning characteristics can be found in Table 3.1. The research presents brain characteristics and its components in Chap. 2. These important concepts can be found in the following.

- **No Invariant Representation**, one thing to one representation, the brain keeps only one thing into one neuron basic. It never changed; even the pattern for that thing changed.
- **No Hierarchy**, many patterns can represent and connect to one thing as it works in a hierarchy. It also helps to reduce storage size or redundancy.
- **No Auto-associated Memory**, even a pattern of one thing is missing. Nevertheless, connections in the hierarchy can help to fill missing patterns. It includes sequential memory as well.
- **No Long-Short Term Memory**, previous information is kept for the short term for human problem-solving. However, they will be promoted to long-term memory if it is the right behavior or rehearsed.
- **No Reward System**, only the right behavior will be learned and selected.
- **No Attention or Focusing**, the brain always predicts all percepts and next actions. Once an unexpected event occurs, attention or focus will be needed. Attention also is used for thinking about each representation.
- **No Filtering**, all precepts are sensed but only focusing will proceed.

Table 3.1 Comparison between methods and learning attributes

Method	Attribute						
	Database	Training data	Deterministic	Manual input	Online	Sequential	Memory
Rule base	N	N	Y	Y	N	N	N
Knowledge base	Y	N	Y	Y	N	N	N
Search	N	N	Y	N	N	N	N
Reinforcement learning	N	N	N	N	Y	N	N
Machine learning	N	Y	N	Y	N	N	N
Deep learning	N	Y	N	Y	N	N	N
Time-dependent learning	N	Y	N	Y	N	Y	Y
HTM	N	Y	N	Y	Y	Y	Y

Table 3.2 Comparison between methods and learning attributes of the brain

Method	Attribute						
	Invariant	Hierarchy	Auto-associated	Short-Long memory	Reward	Attention	Filtering
Rule base	N	N	N	N	N	N	N
Knowledge base	N	N	N	N	N	N	N
Search	N	N	N	N	N	N	N
Reinforcement learning	N	N	N	N	Y	N	N
Machine learning	N	N	N	N	N	N	N
Deep learning	Y	Y	N	N	N	N	N
Time-dependent Learning	N	N	N	Y	N	N	N
HTM	Y	Y	Y	N	N	Y	N

So far, the understanding of the state-of-the-art AI and the functioning of the brain has been elaborated. Comparison between state of the art and attributes of the brain can be summarized and shown in Table 3.2. Many of the techniques discussed in this chapter draw inspiration from certain aspects of the brain, but they still lack a unified conceptual framework. These concepts form the fundamental principles for developing a new algorithm, which will be introduced throughout this book.

3.10 Conclusion

This chapter offers a comprehensive examination of Artificial Intelligence (AI) methodologies, charting their progression from early rule-based systems and knowledge representation to more sophisticated techniques such as machine learning and deep learning. The journey begins with Sect. 3.1, which lays the foundation by defining Artificial Intelligence (AI) and tracing its multidisciplinary roots in philosophy, mathematics, neuroscience, and control theory. It introduces the diverse types of environments that AI agents may encounter, including deterministic, stochastic, fully observable, and partially observable scenarios. This section explains why no single AI approach can address all challenges, highlighting the need for a spectrum of methodologies tailored to different conditions.

Following this, Sect. 3.2 explores knowledge representation, one of the earliest AI approaches rooted in symbolic reasoning. Drawing from propositional and first-order logic, it introduces how inference engines simulate human reasoning by generating conclusions from known facts. While effective in static domains, this section also reveals the inherent limitations of non-learning systems, laying the groundwork for more adaptive methods.

Section 3.3 introduces search algorithms, covering both uninformed and informed search strategies used in deterministic environments, and emphasizes structured decision-making when outcomes are known. Section 3.4 transitioned to machine learning, introducing supervised, unsupervised, and reinforcement learning paradigms. The progression from simple decision trees to ensemble methods like Random Forests and Gradient Boosting illustrated how combining multiple models improves performance. The detailed examination of Support Vector Machines demonstrated how mathematical optimization and kernel methods enable handling of non-linear problems, establishing machine learning as a powerful alternative to rule-based systems.

Building upon this, Sect. 3.5 delves into deep learning, showcasing multilayer neural networks inspired by biological neurons. It emphasizes the scalability and automation of feature learning through architectures such as convolutional neural networks (CNNs). However, the section also cautions against the growing opacity and complexity of these models, which often operate as black boxes and present interpretability challenges.

Section 3.6 addressed sequential learning, progressing from traditional statistical methods like ARIMA to modern neural approaches including RNNs, LSTMs, and GRUs. This evolution demonstrated growing sophistication in handling temporal dependencies, though limitations in long-range dependency modeling and sequential processing bottlenecks remained evident. Section 3.8 introduced the Transformer architecture and Large Language Models, representing a paradigm shift from sequential to parallel processing through attention mechanisms. The detailed exploration of GPT architecture and LLM development phases of pre-training, fine-tuning, and alignment illustrated how modern AI systems achieve remarkable language understanding and generation capabilities while highlighting ongoing challenges in alignment and interpretability.

Section 3.9 presents reinforcement learning as an interactive learning model in which agents improve behavior through feedback in the form of rewards and penalties. This framework echoes behavioral conditioning in biological systems and aligns AI learning with real-world trial-and-error dynamics and long-term policy optimization.

Finally, Sect. 3.10 offers a reflective evaluation of the methodologies introduced throughout the chapter. It acknowledges recurring challenges, including high data dependency, limited contextual understanding, and lack of robustness. The section highlights the ongoing need for more biologically inspired, adaptive, and generalizable AI systems. This perspective not only concludes the chapter but also encourages further exploration of alternative models that draw inspiration from brain architecture.

Building on the diverse AI methods explored in this chapter, Chap. 4 shifts focus to one of the core computational principles: Sparse Distributed Representations (SDRs). Originally introduced in Sect. 2.4, SDRs offer a biologically plausible mechanism for encoding information efficiently and robustly by mimicking the brain's strategy of activating only a small subset of neurons at a time. This concept plays a central role in the following chapters, enabling pattern learning and serving as the foundational memory structure for the proposed method.

References

1. Selman, B., Levesque, H., Mitchell, D.: A New Method for Solving Hard Satisfiability Problems. In: Proceedings of the Tenth National Conference on Artificial Intelligence (AAAI-92), pp. 440–446. AAAI Press, San Jose, CA (1992)
2. Selman, B., Kautz, H., Cohen, B.: Local Search Strategies for Satisfiability Testing. In: Cliques, Coloring, and Satisfiability (DIMACS Series, vol. 26), pp. 521–531. AMS (1996)
3. Davis, M., Logemann, G., Loveland, D.: A Machine Program for Theorem-Proving. Communications of the ACM **5**(7), 394–397 (1962). https://doi.org/10.1145/368273.368557
4. Rautenberg, W.: A Concise Introduction to Mathematical Logic, 3rd edn. Springer, New York (2009)978-1-4419-1220-5
5. Minsky, M.: A Framework for Representing Knowledge. AI Memo 306, MIT Artificial Intelligence Laboratory, Cambridge, MA (1974)
6. van Harmelen, F., Lifschitz, V., Porter, B. (eds.): Handbook of Knowledge Representation. Elsevier, Amsterdam (2007)
7. Elkan, C., Greiner, R.: Building Large Knowledge-Based Systems: Representation and Inference in the Cyc Project. Artificial Intelligence **61**(1), 41–52 (1993). https://doi.org/10.1016/0004-3702(93)90092-P
8. Auer, S., Bizer, C., Kobilarov, G., Lehmann, J., Cyganiak, R., Ives, Z.: DBpedia: A Nucleus for a Web of Open Data. In: The Semantic Web (ISWC 2007). LNCS, vol. 4825, pp. 722–735. Springer (2007). https://doi.org/10.1007/978-3-540-76298-0_52
9. Banko, M., Cafarella, M.J., Soderland, S., Broadhead, M., Etzioni, O.: Open Information Extraction from the Web. In: Proceedings of the 20th International Joint Conference on Artificial Intelligence (IJCAI-07), pp. 2670–2676. Morgan Kaufmann, Hyderabad, India (2007)
10. Chklovski, T., Gil, Y.: An Analysis of Knowledge Collected from Volunteer Contributors. In: Proceedings of the National Conference on Artificial Intelligence, vol. 2, pp. 564–571 (2005)
11. Pellissier Tanon, T., Vrandečić, D., Schaffert, S., Steiner, T., Pintscher, L.: From Freebase to Wikidata. The Great Migration. In: Proceedings of the 25th International Conference on World Wide Web (WWW '16), pp. 1419–1428. ACM, Montreal (2016). https://doi.org/10.1145/2872427.2874809
12. Price, K.: Review of The Handbook of Artificial Intelligence Vol. 1. ACM SIGART Bulletin, (81), 2–3 (1982). https://doi.org/10.1145/1056803.1056807
13. Nilsson, N.J.: Problem-solving Methods in Artificial Intelligence. McGraw-Hill, New York (1971)
14. Hatem, M., Ruml, W.: Simpler Bounded Suboptimal Search. Proceedings of the AAAI Conference on Artificial Intelligence **28**(1), 856–862 (2014). https://doi.org/10.1609/aaai.v28i1.8846
15. Bryce, D., Kambhampati, S., Smith, D.E.: Planning Graph Heuristics for Belief Space Search. Journal of Artificial Intelligence Research **26**, 35–99 (2006). https://doi.org/10.1613/jair.1869
16. Buro, M.: Improving Heuristic Mini-Max Search by Supervised Learning. Artificial Intelligence **134**(1–2), 85–99 (2002). https://doi.org/10.1016/S0004-3702(01)00093-5
17. Ballard, B.W.: The *-Minimax Search Procedure for Trees Containing Chance Nodes. Artificial Intelligence **21**(3), 327–350 (1983). https://doi.org/10.1016/S0004-3702(83)80015-0
18. Quinlan, J.R.: Induction of Decision Trees. Machine Learning **1**(1), 81–106 (1986). https://doi.org/10.1023/A:1022643204877
19. Breiman, L.: Bagging predictors. Technical Report No. 421, Department of Statistics, University of California, Berkeley (1994)

20. Breiman, L.: Random Forests. Machine Learning **45**(1), 5–32 (2001). https://doi.org/10.1023/A:1010933404324
21. Freund, Y., Schapire, R.E.: A decision-theoretic generalization of on-line learning and an application to boosting. In: Proceedings of the Second European Conference on Computational Learning Theory (EuroCOLT), pp. 23–37. Springer (1995)
22. Friedman, J.H.: Greedy Function Approximation: A Gradient Boosting Machine. The Annals of Statistics **29**(5), 1189–1232 (2001). https://doi.org/10.1214/aos/1013203451
23. Meyer, C.D.: Matrix Analysis and Applied Linear Algebra. Society for Industrial and Applied Mathematics, Philadelphia, PA (2000)
24. Altman, N.S.: An Introduction to Kernel and Nearest-Neighbor Nonparametric Regression. The American Statistician **46**(3), 175–185 (1992)
25. Fine, T.L.: Feedforward Neural Network Methodology. Springer, New York (1999)
26. LeCun, Y., Bottou, L., Bengio, Y., Haffner, P.: Gradient-Based Learning Applied to Document Recognition. Proceedings of the IEEE **86**(11), 2278–2324 (1998). https://doi.org/10.1109/5.726791
27. Box, G.E.P., Jenkins, G.M.: Time Series Analysis: Forecasting and Control, 1st edn. Holden-Day, San Francisco (1970)
28. Tsay, R.S.: Analysis of Financial Time Series, 3rd edn. Wiley, Hoboken, NJ (2010). https://doi.org/10.1002/9780470644560
29. Jain, L.C., Medsker, L.R.: Recurrent Neural Networks: Design and Applications. CRC Press, Boca Raton, FL (1999)
30. Hochreiter, S., Schmidhuber, J.: Long Short-Term Memory. Neural Computation **9**(8), 1735–1780 (1997). https://doi.org/10.1162/neco.1997.9.8.1735
31. Cho, K., Merrienboer, B., Gulcehre, C., Bougares, F., Schwenk, H., Bengio, Y.: Learning Phrase Representations using RNN Encoder-Decoder for Statistical Machine Translation. In: Proceedings of the 2014 Conference on Empirical Methods in Natural Language Processing (EMNLP), pp. 1724–1734. ACL, Doha (2014). https://doi.org/10.3115/v1/D14-1179
32. Sutskever, I., Vinyals, O., Le, Q.V.: Sequence to Sequence Learning with Neural Networks. In: Advances in Neural Information Processing Systems (NeurIPS), 27, 3104–3112 (2014). Available: https://arxiv.org/abs/1409.3215
33. Vaswani, A., Shazeer, N., Parmar, N., Uszkoreit, J., Jones, L., Gomez, A.N., Kaiser, L., Polosukhin, I.: Attention is all you need. Advances in Neural Information Processing Systems, 30 (2017)
34. Bahdanau, D., Cho, K., Bengio, Y.: Neural machine translation by jointly learning to align and translate. arXiv preprint arXiv:1409.0473 (2014). Available: https://arxiv.org/abs/1409.0473
35. Devlin, J., Chang, M.-W., Lee, K., Toutanova, K.: BERT: Pre-training of Deep Bidirectional Transformers for Language Understanding. arXiv preprint arXiv:1810.04805 (2018). Available: https://arxiv.org/abs/1810.04805
36. Radford, A., Wu, J., Child, R., Luan, D., Amodei, D., Sutskever, I.: Language models are unsupervised multitask learners. OpenAI Blog, 1(8), 9 (2019). Available: https://cdn.openai.com/better-language-models/language_models_are_unsupervised_multitask_learners.pdf
37. Peters, M.E., Neumann, M., Iyyer, M., Gardner, M., Clark, C., Lee, K., Zettlemoyer, L.: Deep contextualized word representations. arXiv preprint arXiv:1802.05365 (2018). Available: https://arxiv.org/abs/1802.05365
38. Howard, J., Ruder, S.: Universal language model fine-tuning for text classification. arXiv preprint arXiv:1801.06146 (2018). Available: https://arxiv.org/abs/1801.06146
39. Raffel, C., Shazeer, N., Roberts, A., Lee, K., Narang, S., Matena, M., Zhou, Y., Li, W., Liu, P.J.: Exploring the limits of transfer learning with a unified text-to-text transformer. Journal of Machine Learning Research **21**(1), 5485–5551 (2020)
40. Brown, T., Mann, B., Ryder, N., Subbiah, M., Kaplan, J.D., Dhariwal, P., Neelakantan, A., Shyam, P., Sastry, G., Askell, A., et al.: Language models are few-shot learners. Advances in Neural Information Processing Systems **33**, 1877–1901 (2020)
41. Kaplan, J., McCandlish, S., Henighan, T., Brown, T.B., Chess, B., Child, R., Gray, S., Radford, A., Wu, J., Amodei, D.: Scaling laws for neural language models. arXiv preprint arXiv:2001.08361 (2020). Available: https://arxiv.org/abs/2001.08361

42. Lieber, O., Sharir, O., Lenz, B., Shoham, Y.: Jurassic-1: Technical details and evaluation. White paper. AI21 Labs (2021). Available: https://www.ai21.com/research/jurassic-1-technical-details-evaluation/
43. Rae, J.W., Borgeaud, S., Cai, T., Millican, K., Hoffmann, J., Song, F., Aslanides, J., Henderson, S., Ring, R., Young, S., et al.: Scaling language models: Methods, analysis & insights from training Gopher. arXiv preprint arXiv:2112.11446 (2021). Available: https://arxiv.org/abs/2112.11446
44. Smith, S., Patwary, M., Norick, B., LeGresley, P., Rajbhandari, S., Casper, J., Liu, Z., Prabhumoye, S., Zerveas, G., Korthikanti, V., et al.: Using DeepSpeed and Megatron to train Megatron-Turing NLG 530B, a large-scale generative language model. arXiv preprint arXiv:2201.11990 (2022). Available: https://arxiv.org/abs/2201.11990
45. Chen, M., Tworek, J., Jun, H., Yuan, Q., Pinto, H.P.O., Kaplan, J., Edwards, H., Burda, Y., Joseph, N., Brockman, G., et al.: Evaluating large language models trained on code. arXiv preprint arXiv:2107.03374 (2021). Available: https://arxiv.org/abs/2107.03374
46. Thoppilan, R., De Freitas, D., Hall, J., Shazeer, N., Kulshreshtha, A., Cheng, H.-T., Jin, A., Bos, T., Baker, L., Du, Y., et al.: LaMDA: Language models for dialog applications. arXiv preprint arXiv:2201.08239 (2022). Available: https://arxiv.org/abs/2201.08239
47. Hoffmann, J., Borgeaud, S., Mensch, A., Buchatskaya, E., Cai, T., Rutherford, E., Casas, D. de Las, Hendricks, L.A., Welbl, J., Clark, A., et al.: Training compute-optimal large language models. arXiv preprint arXiv:2203.15556 (2022). Available: https://arxiv.org/abs/2203.15556
48. Fedus, W., Zoph, B., Shazeer, N.: Switch Transformer: Scaling to trillion parameter models with simple and efficient sparsity. Journal of Machine Learning Research **23**(1), 5232–5270 (2022)
49. Borgeaud, S., Mensch, A., Hoffmann, J., Cai, T., Rutherford, E., Millican, K., Van Den Driessche, G.B.M., Lespiau, J.-B., Damoc, B., Clark, A., et al.: Improving language models by retrieving from trillions of tokens. International Conference on Machine Learning, 2206–2240 (2022)
50. Chase, H.: LangChain. (2022). Available: https://github.com/hwchase17/langchain [Accessed: 27 May 2025]
51. Ouyang, L., Wu, J., Jiang, X., Almeida, D., Wainwright, C., Mishkin, P., Zhang, C., Agarwal, S., Slama, K., Ray, A., et al.: Training language models to follow instructions with human feedback. Advances in Neural Information Processing Systems **35**, 27730–27744 (2022)
52. OpenAI: Introducing ChatGPT (2022). Available: https://openai.com/index/chatgpt/ [Accessed: 27 May 2025]
53. Touvron, H., Lavril, T., Izacard, G., Martinet, X., Lachaux, M.-A., Lacroix, T., Roziére, B., Goyal, N., Hambro, E., Azhar, F., et al.: Llama: Open and efficient foundation language models. arXiv preprint arXiv:2302.13971 (2023). Available: https://arxiv.org/abs/2302.13971
54. Jiang, A.Q., Sablayrolles, A., Mensch, A., Bamford, C., Chaplot, D.S., Casas, D. de las, Bressand, F., Lengyel, G., Lample, G., Saulnier, L., et al.: Mistral 7B. arXiv preprint arXiv:2310.06825 (2023). Available: https://arxiv.org/abs/2310.06825
55. Sennrich, R., Haddow, B., Birch, A.: Neural machine translation of rare words with subword units. arXiv preprint arXiv:1508.07909 (2015). Available: https://arxiv.org/abs/1508.07909
56. Hendrycks, D., Gimpel, K.: Gaussian error linear units (GELUs). arXiv preprint arXiv:1606.08415 (2016). Available: https://arxiv.org/abs/1606.08415
57. Common Crawl Foundation. Common Crawl. https://commoncrawl.org/ [Accessed: 27 May 2025]
58. Raffel, C., Shazeer, N., Roberts, A., Lee, K., Narang, S., Matena, M., Zhou, Y., Li, W., Liu, P.J.: Colossal Clean Crawled Corpus (C4). 2020. https://huggingface.co/datasets/allenai/c4 [Accessed: 27 May 2025]
59. Gokaslan, A., Cohen, V.: OpenWebText: An open-source replication of OpenAI's WebText. 2024. https://huggingface.co/datasets/Skylion007/openwebtext [Accessed: 27 May 2025]
60. Zhu, Y., Kiros, R., Zemel, R., Salakhutdinov, R., Urtasun, R., Torralba, A., Fidler, S.: Aligning Books and Movies: Towards Story-like Visual Explanations by Watching Movies and Reading Books. arXiv preprint arXiv:1506.06724 (2015). Available: https://arxiv.org/abs/1506.06724 (paper), https://hyper.ai/en/datasets/13642 (dataset). [Accessed: 27 May 2025]

61. Project Gutenberg. Available: https://www.gutenberg.org/ [Accessed: 27 May 2025]
62. Wikimedia Foundation. English Wikipedia Dump. 2025. Available: https://dumps.wikimedia.org/enwiki/ [Accessed: 27 May 2025]
63. Hu, E.J., Shen, Y., Wallis, P., Allen-Zhu, Z., Li, Y., Wang, L., Wang, S., Chen, W., Chen, X., Liu, B., et al.: LoRA: Low-Rank Adaptation of Large Language Models. arXiv preprint arXiv:2106.09685 (2021). Available: https://arxiv.org/abs/2106.09685
64. Banner, R., Nahshan, Y., Hoffer, E., Soudry, D.: Post Training 4-bit Quantization of Convolutional Networks for Rapid Deployment. In: Advances in Neural Information Processing Systems (NeurIPS), 2019. Available: https://arxiv.org/abs/1908.00217
65. Zhao, Y., Chen, X., Wang, Y., Wang, Z., Wei, X., Lin, Y., Zhang, Y., Lin, Z., Han, S.: A Survey of Post-Training Quantization. arXiv preprint arXiv:2405.07135, 2024. Available: https://arxiv.org/abs/2405.07135
66. Dettmers, T., Pagnoni, A., Holtzman, A., Zettlemoyer, L.: QLoRA: Efficient Finetuning of Quantized LLMs. In: Advances in Neural Information Processing Systems (NeurIPS), 2023. Available: https://arxiv.org/abs/2305.14314
67. Gu, Y., Dong, L., Wei, F., Huang, M.: MiniLLM: Knowledge Distillation of Large Language Models. In: International Conference on Learning Representations (ICLR), 2024. Available: https://arxiv.org/abs/2306.08543
68. Rafailov, R., Sharma, A., Mitchell, E., Ermon, S., Manning, C.D., Finn, C.: Direct Preference Optimization: Your Language Model is Secretly a Reward Model. In: Advances in Neural Information Processing Systems (NeurIPS), 2023. Available: https://arxiv.org/abs/2305.18290
69. Puterman, M.L., Shin, M.C.: Action Elimination Procedures for Modified Policy Iteration Algorithms. Operations Research **30**(2), 301–318 (1982)
70. Bellman, R.: Dynamic Programming and Stochastic Control Processes. Information and Control **1**(3), 228–239 (1958). https://doi.org/10.1016/S0019-9958(58)80003-0
71. Tassa, Y., Erez, T., Todorov, E.: Synthesis and Stabilization of Complex Behaviors Through Online Trajectory Optimization. In: 2012 IEEE/RSJ International Conference on Intelligent Robots and Systems (IROS), pp. 4906–4913. IEEE (2012). https://doi.org/10.1109/IROS.2012.6386025

Chapter 4
Spare Distributed Representation (SDRs)

In this chapter builds upon the foundational understanding of the brain's architecture explored in Chaps. 1 and 2 and the diverse Artificial Intelligence (AI) approaches examined in Chap. 3. The previous chapters introduced core biological principles such as sparsity, hierarchy, and plasticity in the neocortex, highlighting how the brain efficiently processes and stores information. Meanwhile, Chap. 3 surveyed a range of AI techniques, tracing their evolution, strengths, and limitations, and ultimately revealing the need for more adaptive, robust, and biologically grounded models. Chapter 4 introduces Sparse Distributed Representations (SDRs) as a concrete, biologically inspired computational mechanism that embodies these neural principles.

SDRs provide a mathematically rigorous and computationally efficient way to encode information by mimicking how the brain uses sparse patterns of neural activity to process and store knowledge. Unlike dense representations used in traditional computing and deep learning, SDRs are robust, noise-tolerant, and highly energy-efficient. This chapter explains how SDRs allow systems to learn, generalize, and recognize patterns in a way that mirrors cognitive processing, bridging biological plausibility with computational practicality.

In this chapter begins with an introduction to the biological foundations of Sparse Distributed Representations (SDRs) (Sect. 4.1), emphasizing how sparsity, invariance, and hierarchical organization in the neocortex inspire this efficient and robust encoding scheme. Section 4.2 defines SDRs as high-dimensional binary vectors with a small percentage of active bits, allowing meaningful and fault-tolerant representations of complex concepts.

Section 4.3 describes how semantic similarity is determined by the overlap between active bits, enabling inexact matching and generalization in noisy environments. Section 4.4 examines SDR capacity and false positives, showing how sparse structures can represent vast numbers of patterns with minimal collision risk, even under relaxed thresholds. Section 4.5 introduces subsampling, a technique that retains

T. Soisoonthorn and H. Unger, *A Brain-Inspired Approach to Natural Language Processing*, Studies in Computational Intelligence 1225,
https://doi.org/10.1007/978-3-032-00014-9_4

discriminative power while using only partial SDRs, making it ideal for resource-constrained systems. Section 4.6 presents the union operation, which merges multiple SDRs for efficient set representation and fast membership checks, while balancing false positive risks through parameter tuning.

Section 4.7 highlights the computational and energy efficiency of SDRs, which leverage bitwise operations and sparse data structures to achieve fast, low-power processing on standard hardware. Finally, Sect. 4.8 provides the chapter's conclusion.

4.1 Introduction

Understanding how the human brain, especially the neocortex, processes information offers profound insights for designing intelligent systems. Among the most striking principles observed in neuroscience is the concept of sparse activation. At any given moment, only a small percentage of neurons in the brain are active, while the vast majority remain silent. This is not a limitation; it is an advantage. Sparsity helps the brain conserve energy and improves robustness, allowing it to operate efficiently even in the presence of noise or uncertainty. Experimental evidence [1, 2] from sensory and motor areas confirms that the brain uses sparse neural activity to encode information in a reliable and efficient manner.

Another key principle is invariant representation, which refers to the brain's ability to recognize the same object or concept despite changes in perception. For example, a neuron may respond to a familiar face regardless of the viewing angle or lighting. This reflects the brain's capacity to capture meaning beyond surface details. In addition to stability, the brain's representations are hierarchical. It does not process entire objects at once. Instead, it builds them up from simpler components. Lower levels of the neocortex detect basic features such as edges, colors, or small shapes. These are progressively combined at higher levels into more complex representations, such as entire objects, categories, or abstract concepts. This layered structure allows for generalization and flexible understanding, even when encountering unfamiliar or partially obscured inputs.

Supporting all this is the uniform architecture of the neocortex. Despite handling different functions such as vision, hearing, and movement, its structure remains largely the same across regions. Each area consists of repeating units called cortical columns, which can perform similar types of processing. This modularity and consistency allow different regions to compensate for one another. For instance, after a stroke, one part of the brain can often relearn lost functions using other regions. This phenomenon is known as neural plasticity.

The biological principles of sparsity, invariance, hierarchy, and uniformity form the foundation for a powerful brain-inspired framework known as Sparse Distributed Representations (SDRs) [3, 4]. SDRs are used in computational models like Hierarchical Temporal Memory (HTM) to emulate the way the brain encodes, processes, and understands information.

4.2 SDR Vectors

An SDR (Sparse Distributed Representation) is a binary vector composed of many bits, with only a small percentage of them active at any given time. This characteristic makes it sparse. Each bit in the vector corresponds conceptually to a neuron. A "1" indicates activation, while a "0" means inactivity. Critically, each active bit has semantic meaning, and the combination of active bits represents a specific concept or pattern. When two SDRs share active bits in the same positions, they share semantic features. For example, a shared active bit might represent the attribute "red", meaning both vectors encode that feature. By measuring how many bits overlap, we can instantly assess the degree of similarity between two representations. This overlap mechanism allows systems based on SDRs to generalize meaningfully, just as the brain does.

Because of their structure, SDRs can represent virtually anything, from sensory inputs like images and sounds to abstract ideas like words, locations, or behaviors. This makes them incredibly flexible. In fact, much like the neocortex, systems can process any kind of data as long as it is encoded as an SDR. The algorithms themselves do not need to "know" what the bits represent. Instead, the intelligence is embedded in the data representation, not in the code.

This approach stands in stark contrast to how traditional computers represent information. Conventional systems rely on dense representations. For instance, the ASCII code for the letter "O" is 01101111, where each bit contributes to the whole meaning but has no meaning on its own. If one bit changes, the entire interpretation can shift–such as changing "O" to a completely different letter. Dense encodings are therefore fragile and inflexible.

By comparison, SDRs are resilient and robust. Even if a few active bits are altered, the core meaning often remains intact. For example, a 2,000-bit SDR with only 40 bits active (2%) can represent concepts using combinations like "red," "leg," "noise," or "eye." If two SDRs share even a few active bits, we can infer shared meaning with high confidence.

Definition 4.1 A *Sparse Distributed Representation (SDR)* is a binary vector in n dimensions, denoted as $x = [b_0, b_1, \ldots, b_{n-1}]$, where each $b_i \in \{0, 1\}$. The number of bits with value 1 is denoted by w_x, also referred to as the active bits. The overlap between two SDR vectors refers to the number of positions in which both vectors have a 1, and it serves as a measure of similarity.

For example:

$$X = [0000010101000000000000101010000000000000]$$

$$Y = [0001000101000000100000010100000000000000]$$

The vectors X and Y have $n = 40$, $w = 6$, overlap $= 4$, and sparsity $s = \frac{w}{n} = \frac{6}{40} = 15\%$.

4.3 Similarity

In the context of Sparse Distributed Representations (SDRs), overlap refers to the number of active bits that are shared between two SDR vectors at the same positions. Each SDR is a binary vector, typically with a large number of bits (e.g., 2048), where only a small subset of bits are active. To determine how semantically similar two SDRs are, we compute the overlap by counting how many bits are set to 1 in both vectors simultaneously. Mathematically, this is defined as the dot product of two binary vectors:

$$\text{Overlap}(A, B) = \sum_{i=1}^{n} A_i \cdot B_i \tag{4.1}$$

where A_i and B_i are the i-th bits of vectors A and B. This operation can also be computed efficiently using a bitwise AND followed by counting the number of resulting 1 s:

$$\text{Overlap}(A, B) = \text{Count}(A \text{ AND } B) \tag{4.2}$$

In practical systems, a threshold θ is applied to determine whether two SDRs are considered a match. If the overlap between two SDRs is greater than or equal to this threshold, the vectors are said to match semantically:

$$\text{Match}(A, B) \iff \text{Overlap}(A, B) \geq \theta \tag{4.3}$$

This threshold allows for *inexact matching*, which makes SDR systems robust to noise and partial corruption. Two SDRs with a high overlap are considered semantically similar, even if they are not identical. For example, consider two 16-bit SDRs: $A = 0101010001000000$ and $B = 1100010001000000$. Performing a bitwise AND results in 0100010001000000, which contains three bits set to 1. If the threshold $\theta = 3$, then these SDRs would be considered a match. This simple yet powerful mechanism supports robust pattern recognition and generalization, making SDRs ideal for systems that require noise tolerance and semantic similarity detection.

4.4 Capacity and False Positives

Sparse Distributed Representations (SDRs) are designed to encode information with both efficiency and robustness. Understanding how many patterns can be reliably represented and distinguished in an SDR system is essential for designing models that are both scalable and tolerant to noise. One of the fundamental properties of SDRs is their ability to uniquely represent patterns using a sparse set of active bits in a high-dimensional binary vector. The total number of distinct patterns that can be represented depends on how strictly we define a match between two SDRs.

In the case of **exact matching**, two SDRs are considered to match only when all w active bits occur in the exact same positions. If we require two SDRs to have an exact match (i.e., $\theta = w$), then even a single bit of noise or corruption in either SDR would result in a mismatch, causing a false negative. In this strict matching scheme, the representational capacity is at its maximum because no patterns overlap. To calculate how many such unique patterns exist, we use the binomial coefficient:

$$\text{Capacity}_{\text{exact}} = \binom{n}{w} = \frac{n!}{w!(n-w)!} \tag{4.4}$$

This formula counts the number of distinct ways to choose w active bits out of n possible positions. For instance, if $n = 2048$ and sparsity is 10% (i.e., $w = 204$), the number of unique combinations is approximately 6.99×10^{286}. Even with lower sparsity, such as $w = 40$, the number of possible SDRs remains extremely large, around 2.37×10^{84}.

It is worth noting that this number is significantly smaller than the total number of encodings possible with dense binary vectors, which is 2^n for a vector of length n. At first glance, this may appear to be a loss in capacity, since dense vectors can theoretically represent more patterns. However, this theoretical reduction is practically irrelevant. SDRs provide more than enough capacity for real-world applications, while offering substantial advantages in robustness, generalization, and computational efficiency. Their sparsity ensures that even at high capacity, patterns do not interfere with one another, enabling fault-tolerant and semantically meaningful representations. This makes SDRs particularly well-suited for brain-inspired learning systems.

An important aspect of capacity is the likelihood of accidental collisions, which refers to two different inputs generating the exact same SDR. If such a collision occurs, the system loses the ability to distinguish between them, leading to confusion or incorrect behavior, similar to assigning the same ID to two different individuals. The probability of this happening is the reciprocal of the total number of unique SDRs:

$$P(\text{match}) = \frac{1}{\binom{n}{w}} \tag{4.5}$$

This expression tells us how likely it is that two randomly generated SDRs with the same parameters (n and w) will be identical. Because the number of possible SDRs grows combinatorially, this probability becomes vanishingly small in practice. For example, with $n = 2048$ and $w = 40$, the chance of such a random collision is essentially zero (4.21×10^{-85}). This ensures that SDR-based systems can scale to represent vast numbers of distinct patterns reliably and without ambiguity.

While exact matching provides the highest possible capacity, its intolerance to any variation makes it impractical for most real-world applications. To improve robustness in practical systems, we typically use **inexact matching**, where two SDRs are considered a match if they share at least θ overlapping active bits, where $\theta \leq w$. This allows systems to tolerate minor distortions or noise in the input, making them more resilient under real-world conditions.

Formally, the inexact matching rule is defined as:

$$\text{Match}(A, B) \iff \text{Overlap}(A, B) \geq \theta \tag{4.6}$$

This flexibility introduces the possibility of *false positives*, which are cases where two unrelated SDRs are mistakenly considered similar because they happen to overlap in θ or more positions purely by chance. To evaluate how likely this is, we calculate the probability that two randomly generated SDRs with the same parameters (n, w) will overlap in at least θ positions. The probability of a false positive is given by:

$$P_{\text{false positive}} = \frac{\sum_{b=\theta}^{w} \binom{w}{b} \cdot \binom{n-w}{w-b}}{\binom{n}{w}} \tag{4.7}$$

While the formula may appear complex, the result is often reassuring: for typical values (e.g., $n = 2048$, $w = 40$, $\theta = 20$), the probability of a false match is only 2.49×10^{-26}. This low false positive rate makes inexact matching both robust and reliable.

To understand how inexact matching affects capacity, we define the *effective capacity* as the number of distinct patterns a system can safely encode while allowing inexact matching with a given threshold θ. Since many SDRs will fall within the same match "volume", the effective capacity is lower than in exact matching and is calculated as:

$$\text{Capacity}_{\text{inexact}} = \frac{\binom{n}{w}}{V(n, w, \theta)} \tag{4.8}$$

Here, $V(n, w, \theta)$ is the number of SDRs that overlap with a given SDR in at least θ positions. This matching volume is computed as:

$$V(n, w, \theta) = \sum_{b=\theta}^{w} \binom{w}{b} \cdot \binom{n-w}{w-b} \tag{4.9}$$

By dividing the total number of SDRs by the number of possible matches per SDR, we get a realistic estimate of how many distinguishable patterns the system can represent while allowing for inexact matching. For example, when $n = 2048$, $w = 40$, and $\theta = 20$, the effective capacity is approximately 4.01×10^{25}. This allows the system to reliably represent trillions of distinct patterns without confusion. This balance between capacity and tolerance is one of the key strengths of SDRs: they support both efficient memory usage and noise-robust pattern recognition, making them ideally suited for brain-inspired computational models.

4.5 Subsampling

In practical applications, it is often not feasible to store or process the entire high-dimensional representation of an SDR due to constraints such as limited memory, communication bandwidth, or processing speed. In such cases, a common solution is to use only part of the full SDR. This approach is called *subsampling*, and it involves selecting a subset of the bits in the SDR to use for comparison or classification tasks.

An SDR is typically a binary vector of length n, with exactly w active bits. When subsampling is applied, only n' bits are selected from the full vector, where $n' < n$. Depending on which bits are chosen, the subsampled vector, denoted as x', may still contain some of the original active bits. We can then compare x' to another SDR y to determine if they match, using a threshold θ' that specifies the minimum number of overlapping active bits required.

To evaluate how reliable this matching is, we calculate the probability that a randomly generated SDR y will falsely appear to match the subsampled SDR x' by chance. This false match probability is given by:

$$P_{\text{false match}} = \frac{\sum_{b=\theta'}^{w'} \binom{w'}{b} \cdot \binom{n'-w'}{w'-b}}{\binom{n'}{w'}} \tag{4.10}$$

Here, w' is the number of active bits in the subsampled SDR x', and b is the number of overlapping bits between x' and y. The numerator counts all the possible SDRs that overlap with x' in at least θ' positions, while the denominator gives the total number of possible SDRs of the same size and sparsity. This probability decreases exponentially as n', w', and θ' increase, which means that even with partial information, SDRs can maintain strong resistance to false matches.

For instance, suppose we begin with a subsampled SDR of size $n' = 512$ and with $w' = 20$ active bits. If we set the matching threshold to $\theta' = 10$, the probability that a randomly generated SDR falsely matches the subsampled pattern drops to approximately 9.09×10^{-11}. This result is striking, considering we are only using a partial view of the full SDR, and yet the system retains strong discriminative power. Now imagine we double the size of the subsampled vector to $n' = 1024$, keeping the number of active bits proportionally increased to $w' = 40$, and set the threshold to $\theta' = 20$. In this case, the probability of a false positive becomes so low, far below one in 2.36×10^{-20}, that it is effectively negligible for most practical purposes.

These results demonstrate that subsampling can preserve the powerful discriminative properties of SDRs, even when using only a fraction of the original vector. When parameters are selected appropriately, this enables efficient, accurate matching in resource-constrained environments.

Subsampling, however, introduces a trade-off. Reducing the number of bits lowers memory usage and speeds up computation but can also weaken the system's ability to distinguish between patterns. If too few active bits are preserved, or the matching threshold is too low, the risk of false positives increases.

Despite this, subsampling remains highly valuable in settings such as embedded devices, distributed sensor networks, and real-time inference systems. It enables fast and efficient pattern recognition using limited data. Subsampling is also beneficial after union operations, which increase bit density; applying subsampling afterward can restore sparsity while still allowing approximate matching.

4.6 Union

A key property of Sparse Distributed Representations (SDRs) is their ability to represent multiple distinct patterns using a single composite vector. This is achieved through an operation called the *union*. In SDRs, a union is created by performing a bitwise OR across two or more SDRs. The result is a new SDR in which each bit is set to 1 if it was active in any of the input vectors. Formally, given SDRs A, B, and C, the union U is defined as:

$$U = A \vee B \vee C$$

where $\vee$ denotes the bitwise OR operation. This process can be extended to combine any number of SDRs. For example, suppose we have three SDRs of length 6:

$$A = [0, 0, 1, 0, 0, 1], \quad B = [0, 0, 1, 1, 0, 0], \quad C = [1, 0, 0, 0, 1, 0]$$

We can compute the union step by step:

$$U = A \vee B \vee C = [1, 0, 1, 1, 1, 1]$$

Now, suppose we want to check whether A matches U. We compute the overlap between A and U, which is simply the number of positions where both vectors have 1 s. In this case:

$$\text{Overlap}(A, U) = 2$$

Since all the 1 s in A are present in U, we can say that A is still represented in the union. This principle allows for fast membership testing. If an input SDR shares enough overlapping bits with a union, it is assumed to be part of the set.

It is important to note that the union operation is not reversible. Once multiple SDRs are merged, there is no way to separate them back into their original components. Given the final union $U = [1, 0, 1, 1, 1, 1]$, we cannot tell which exact SDRs contributed to which bits. This trade-off is accepted in exchange for memory efficiency and fast lookup.

While unioning is efficient for representing a set of SDRs, it increases the number of active bits in the result. As more SDRs are added, the vector becomes denser, which increases the chance of false positives, where unrelated SDRs match the union by coincidence. To quantify this risk, we apply the probabilistic model described by Ahmad and Hawkins [3] for estimating the false positive rate under exact matching conditions. Given n total bits per SDR and w active bits per SDR, the sparsity is defined as $s = \frac{w}{n}$. When M SDRs are unioned, the probability that a given bit remains "0" after all unions is:

$$p_0 = (1 - s)^M \tag{4.11}$$

The probability that a bit is "1" in the union is:

$$1 - p_0 = 1 - (1 - s)^M \tag{4.12}$$

A false positive occurs when a new, random SDR with w active bits happens to fall entirely within the set bits of the union. The probability of this happening, under exact match conditions where all w bits must overlap, is:

$$p_{\text{fp}} = \left(1 - (1 - s)^M\right)^w \tag{4.13}$$

To illustrate this, consider SDRs of length $n = 2048$ with $w = 40$ active bits. If we union together $M = 10$ such SDRs, then fp is 1.30×10^{-30}. This means that even after storing 10 SDRs in the union, the chance that a random SDR falsely matches is very low. This robustness is a key advantage of SDRs, even when performing aggressive memory compression.

The previous analysis of union in SDRs assumed perfect, exact matches, where the threshold for a match θ is equal to the number of active bits w. However, real-world data is often noisy or incomplete, so it is more practical to relax this condition and allow matches to occur when the overlap between an input and the unioned SDR exceeds some threshold $\theta < w$. This shift from exact to inexact matching leads to a more realistic and robust framework for understanding how SDRs behave under imperfect conditions.

In inexact matching, we can no longer rely on deterministic counts of 1-bits in the union. Instead, we estimate the *expected number of 1 s* in the unioned vector X, which is given by:

$$\hat{w}_X = n(1 - p_0), \quad \text{where } p_0 = (1 - s)^M \text{ and } s = \frac{w}{n} \tag{4.14}$$

Here, M is the number of SDRs that have been unioned, n is the total length of the SDR vector, and s is the sparsity. This expectation provides a practical estimate of how many positions in the unioned vector are likely to be 1, without requiring explicit tracking of each bit.

To assess the likelihood of a random SDR y overlapping with X, we calculate the expected number of matching cases where exactly b of the w bits in y align with 1-bits in the union:

$$E[|\Omega_X(n, w, b)|] = \binom{\hat{w}_X}{b} \cdot \binom{n - \hat{w}_X}{w - b} \tag{4.15}$$

This expression represents the number of possible input SDRs with exactly b overlapping bits. To compute the *false positive rate* for inexact matches, we sum over all values from $b = \theta$ to w:

$$\varepsilon \approx \frac{\sum_{b=\theta}^{w} |\Omega_X(n, w, b)|}{\binom{n}{w}} \tag{4.16}$$

This formulation estimates the chance that a new, unrelated input matches the unioned SDR by random overlap. In practice, this approximation is effective and computationally tractable. Moreover, in many cases, the result is dominated by one term (usually where $b = \theta$), simplifying the expression to:

$$\varepsilon \approx \frac{|\Omega_X(n, w, \theta)|}{\binom{n}{w}} \quad \text{(dominant term approximation)} \tag{4.17}$$

Understanding this trade-off is essential. As expected, lowering the threshold θ increases the risk of false matches. However, this risk can be mitigated by increasing the SDR size n, which reduces the overall density of 1 s, or by selecting a more conservative (higher) threshold to require stronger overlap for a match.

For instance, consider an SDR of length $n = 2048$ with $w = 40$ active bits, and a union formed from $M = 40$ such SDRs. Under *exact matching* conditions ($\theta = w = 40$), the false positive rate is 2.13×10^{-11}. However, if we relax the match threshold to $\theta = 30$, the false positive rate increases to 3.81×10^{-3}.

To mitigate this, we can increase the SDR length to $n = 4096$, keeping $w = 40$ and $\theta = 30$. In this case, the lower sparsity significantly reduces the chance of accidental overlap, and the false positive rate drops dramatically to 3.11×10^{-8}.

In summary, the union operation allows SDRs to efficiently represent multiple patterns in a single composite vector by performing a bitwise OR. While exact matching ensures extremely low false positive rates, it is often too rigid for noisy or incomplete data. In practice, relaxing the threshold to allow inexact matching, where an input is considered a match if it overlaps with the union in at least $\theta < w$ positions, offers a more robust and scalable alternative. Results show that as θ decreases or the number of unioned SDRs increases, false positive rates rise. However, this can be counterbalanced by increasing the SDR size n. Through careful tuning of n, w, M, and θ, SDR-based systems can maintain both compactness and high reliability, even when aggressively compressing memory representations.

4.7 Computation and Energy Efficiency

From a practical standpoint, Sparse Distributed Representations (SDRs) are highly efficient in terms of both computation and energy. SDRs are implemented as large binary vectors with only a small percentage of bits active at any given time. This sparsity, combined with their binary nature, enables extremely fast and lightweight processing on standard hardware.

Because SDRs use only 0 s and 1 s, they can be manipulated using simple bitwise operations such as AND, OR, and XOR. These operations are natively supported by CPUs and require far fewer computational resources than floating-point arithmetic or dense matrix operations commonly used in neural networks. As a result, SDRs do not require specialized hardware such as GPUs or TPUs, and can be deployed effectively on low-power devices, edge systems, and embedded platforms.

Another important efficiency gain comes from how SDRs are stored and compared. Since only a small number of bits (typically 1–5%) are active, there is no need to store the entire vector. Instead, we can represent an SDR by storing the indices of the active bits, dramatically reducing the memory footprint. For instance, a 2048-bit SDR with 40 active bits can be stored in just 40 integers, rather than 2048 bits. Similarity comparison between two SDRs then becomes a matter of counting overlapping indices, which is computationally trivial.

This sparse and index-based representation also improves energy efficiency. By activating only a small subset of computational paths, SDR-based systems inherently reduce power consumption, much like how the brain minimizes metabolic cost by limiting neural activity. As such, SDRs offer a biologically inspired, hardware-friendly alternative to traditional dense representations, supporting both high scalability and energy-conscious computing.

4.8 Conclusion

This chapter has introduced Sparse Distributed Representations (SDRs) as a biologically inspired mechanism for encoding information, grounded in principles observed in the neocortex. The chapter begins with Sect. 4.1, which establishes the neuroscience foundation that motivates SDRs as a model for efficient and robust information processing. Understanding this biological basis is essential, as it informs the design of representations that mirror how the brain encodes meaning with minimal activity.

Section 4.2 builds on this by defining the structure of SDRs as high-dimensional binary vectors with sparse activations and explaining their core properties. This sets the stage for Sect. 4.3, which introduces overlap-based similarity as a way to determine semantic closeness between SDRs. This concept is critical for enabling inexact matching and generalization, making SDRs suitable for real-world, noisy data.

To evaluate the scalability of SDRs, Sect. 4.4 examines their capacity and false positive rates, showing how sparse encoding supports a vast representational space with low ambiguity. Section 4.5 extends this by exploring subsampling, which maintains recognition accuracy even when using only parts of the original SDR. This enables practical use in resource-constrained environments.

Section 4.6 introduces the union operation, allowing multiple SDRs to be combined into a single representation for set membership testing and efficient storage. This builds upon earlier concepts by leveraging sparsity and similarity to support scalable memory compression. Finally, Sect. 4.7 addresses the computational and energy efficiency of SDRs, demonstrating how their binary and sparse nature supports fast, low-power processing without specialized hardware.

These concepts lay the foundation for Chap. 5, which shifts the focus from static representations to dynamic sequence learning. While SDRs define what is represented, Chap. 5 explores how information unfolds over time. This is a critical function of cognition. It introduces a brain-inspired sequence memory model that emulates the brain's use of short-term and long-term memory to learn, retain, and recall temporal patterns such as language, transactions, and behaviors. Built upon the SDR framework, the model incorporates auto-association, prediction, filtering, and forgetting. These mechanisms enable real-time learning and decision-making. The approach is further validated through a hardware simulation, demonstrating its practical implementation and efficiency. Its application is not limited to hardware environments.

References

1. Hromádka, T., DeWeese, M.R., Zador, A.M.: Sparse representation of sounds in the unanesthetized auditory cortex. PLoS Biol. **6**(1), e16 (2008). https://doi.org/10.1371/journal.pbio.0060016
2. Churchland, M.M., Cunningham, J.P., Kaufman, M.T., Ryu, S.I., Shenoy, K.V.: Cortical preparatory activity: representation of movement or first cog in a dynamical machine? Neuron **68**(3), 387–400 (2010). https://doi.org/10.1016/j.neuron.2010.09.015
3. Ahmad, S., Hawkins, J.: Properties of Sparse Distributed Representations and their Application to Hierarchical Temporal Memory (2015). arXiv:1503.07469
4. Hawkins, J., Ahmad, S., Purdy, S., Lavin, A.: Biological and Machine Intelligence (Version 0.4). Numenta (2016). https://numenta.com/resources/biological-and-machine-intelligence/. Accessed 4 May 2025

Chapter 5
A New Brain-Inspired Sequence Learning Memory

While understanding static representations in Chap. 4 is an important step, it may not be sufficient on its own to capture the complexity of real-world cognition, which often involves patterns that unfold over time. Everyday tasks, such as interpreting language, making decisions, or responding to changing environments, require the ability to process sequences. In this context, Chap. 5 explores the temporal aspect of learning by proposing a memory model inspired by how the brain may handle sequences through mechanisms such as association, forgetting, and prediction.

Rather than relying on complex or resource-intensive training methods, this chapter introduces a lightweight sequence learning approach based on Sparse Distributed Representations (SDRs). The method is intentionally simple, inspired by the brain's own mechanisms. It models the dynamic interaction between short-term and long-term memory in a simplified yet biologically plausible way. To demonstrate its feasibility and efficiency, the model is implemented through a hardware simulation using generic, low-cost components. This hardware-based demonstration not only validates the core concepts in a tangible way but also suggests potential for real-time, embedded systems that require low energy consumption and fast response, which is particularly valuable in edge computing and constrained environments.

In this chapter opens with a discussion on how the brain processes sequences of information over time (Sect. 5.1), highlighting mechanisms such as auto-association, filtering, and memory consolidation. It introduces the idea that frequent exposure strengthens memory retention, while rare or noisy inputs are filtered out, mirroring how the brain balances learning and forgetting. Section 5.2 then outlines a conceptual design that models this process using shift registers and counters to manage sequence formation and promotion between memory stages.

Section 5.3 explains how SDRs are structured hierarchically to represent characters, words, and sentences using a hash-based encoding. Section 5.4 adds fault tolerance through logical bitwise operations (e.g., AND, XOR), allowing the system to detect noise or recognize similarity between sequences without heavy

T. Soisoonthorn and H. Unger, *A Brain-Inspired Approach to Natural Language Processing*, Studies in Computational Intelligence 1225,
https://doi.org/10.1007/978-3-032-00014-9_5

computation. Section 5.5 applies these mechanisms to NLP pattern matching, showing how the system predicts the next element in a sequence or enters learning mode when encountering unfamiliar patterns.

Section 5.6 compares this approach to deep learning, noting its advantages in interpretability, data efficiency, and online learning capabilities. Finally, Sect. 5.7 presents a hardware simulation built using MATLAB Simulink, demonstrating that the model can be implemented on low-cost, generic components to support real-time learning and inference in constrained environments. Section 5.8 concludes the chapter.

5.1 Sequence Learning Concept

In Sect. 2.2, it is mentioned that the brain handles input patterns that evolve over time, known as temporal patterns. These patterns are captured as a sequence of representations and processed sequentially. Many real-world datasets also exhibit such temporal characteristics and ordering, for example, as Fig. 5.1, DNA and RNA, e-Learning, web click streaming, text, and purchase transactions. This problem resembles Sequential Pattern Mining [1] that discovers interesting sub-sequences in a set of sequences, except this chapter is for a learning model.

The brain processes sequential learning by receiving a series of inputs in succession. Initially, the brain relies on *short-term memory*, *long-term memory*, and *working memory* to store these representations. When the first representation, *A*, is

CTAGGATTCAGTACGTTGCACTGACTTAGCCTGAGTGCTGTAGTCGAA
TACCTGTTGAGTATGGCACCGGATACCGTGAGGTTTAACTACCGTGAAG
GAGGCTTATCTAACCGTGACCTGAAACCTGTGAACCTGTCGCTTGTAGG

MTVQKNYLSAESVARVQRLIDYYRGSFHLGWFYDREEDVKLEGIRHFV
KELAEEKRAAAEKLLHMRNNRAGRTLFQDLQKPGQDEVGGTPDALKAV
IILDRSINGTLLDVHGLASAQAEPHVLCDYLKSHFVEEEVSLIKLADHLVS
IHQLVGTHAGRGEFLHERLTLKHID

⟨301002, 2025-03-14, 09:30 AM, MetroCenter, {BrownBread, OrangeJuice}⟩
⟨302203, 2025-03-14, 12:45 PM, MetroCenter, {ChickenWrap, Cola, Pear}⟩
⟨301002, 2025-03-14, 05:10 PM, QuickMart, {SoyMilk, Oats, Greens}⟩

Fig. 5.1 DNA, RNA, and purchase transaction data

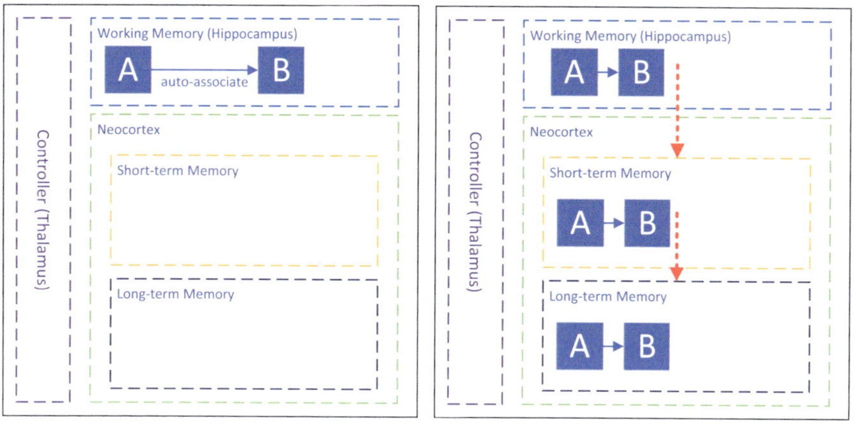

Fig. 5.2 The concept of learning sequence to move information from the working memory to the short-term memory and the long-term memory

received, it is stored in the working memory as a neural unit. Upon recalling step *A*, this representation is activated. The brain connects the first and second representations through auto-association when the second representation, *B*, is received, as shown in Fig. 5.2a. By recalling something from one representation, such as step *A*, the brain automatically recalls the next step, *B*, which facilitates learning the sequential relationship between the two steps. This process is repeated for subsequent steps until the task is completed. Although initially challenging to remember step *A*, once it is recalled, the brain can easily remember step *B*.

Each neural unit in working memory represents what is currently being focused on and thought about, and can be activated sequentially and irreversibly. When the first step is received and stored in a neural unit, it automatically connects to the next neural unit in the sequence to receive the following input. Once the task is finished, these sequences of neural units are transferred to the memory for storage.

Even though the ability to remember a sequence is present, there may still be ambiguity about its significance and whether it should be learned. To ascertain its importance, filters are required to determine if this information is critical enough to keep. The concept is simple: if it's noise, it should not appear frequently; if it's a mistake, it should show up randomly and not recur. Two systems help facilitate this process.

The first one involves shifting from short-term to long-term memory. This is achieved by temporarily housing the sequence (A->B) in short-term memory and using a trial and error method to evaluate if the sequence can fulfill its intended purpose or if it is encountered frequently. If the sequence successfully passes this assessment, it is upgraded to long-term memory, as depicted in Fig. 5.2b. Hence, it can be confirmed that the sequence has been effectively learned. In other words, the information is temporarily stored in the short-term memory, but it can't be employed as its credibility hasn't been established until it gets promoted to long-term memory.

The second system demonstrates that information in the long-short term memory can be discarded over time. Forgetting is an active, necessary process that enables the brain to filter out irrelevant information and prioritize what is most useful for future tasks. However, the time or frequency of exposure to the information also matters. When a human encounters something for the first time, it is easy to forget. However, if the same thing is seen a second time, it becomes harder to forget. In the brain, if we are repeatedly exposed to certain information, the connection will be strengthened, making it harder to forget. In other words, information in the long-short term memory can be forgotten over time, but frequent exposure makes the information in long-term memory or what is seen more often harder to forget compared to short-term memory or less frequently encountered information.

To represent sequential data, we can use a sequence table that contains multiple representations in a sequence. A set of representations, $R = \{r_1, r_2, r_3, ..., r_m\}$, is defined, and a subset of R is a representation set X, denoted as $X \subseteq R$. The size of X is $|X|$, which indicates the number of representations in X. The frequency of occurrence of a sequence is denoted as $Count$, similar to the concept of $frequency$ in [2]. The $Time$ column represents the process of forgetting or decrementing over time, with a higher $Count$ value implying a larger $Time$ value to be deducted. Each representation in a sequence is automatically associated from left to right. Table 5.1 shows the sequence table that contains the sequences and their counts that the brain keeps.

Moreover, the ability to rapidly process interactions with the world is crucial, and it is impractical to continually learn and process every step involved in thinking, information processing, and decision making. Instead, because we have already learned sequences that are stored in long-term memory, the brain can use this information to *process* and *predict* the next step immediately.

When the brain's predictions fail, consciousness arises, prompting a switch to learning mode. This is referred to as *online learning*, enabling adaptation to new and unexpected situations. However, the input representation may sometimes contain noise, making it challenging to differentiate between a sequence that has already been learned and one that requires learning. To overcome this, comparisons are made between the long-term and working memory sequences to determine any differences. If the input representation does not correspond to a prediction, a switch to learning

Table 5.1 Sequence table

Sequence ID	Sequence	Count	Time
1	$\{B, U, T\}$	3	300
2	$\{B, A, L, L\}$	4	400
3	$\{E, A, T\}$	5	500
4	$\{B, A, T\}$	1	50

mode is triggered and any decision-making is postponed until the task is completed. If it is determined that the input is noise, a correction is made. However, if it is a new sequence, it is stored in long-term memory".

The study proposes a novel mechanism for comparing sequences in working memory and long-term memory. It involves comparing each neural unit in the working memory to each neural unit in the long-term memory, starting from the beginning and going all the way to the end. If there are only a few different representations in the sequence, it is likely that it is either noise or a previously learned sequence. However, if there are many different representations, it is likely a new and distinct sequence.

When encountering a new task, the brain utilizes the first input representation (e.g., step A) to search for related sequences in long-term memory. Drawing on past experiences, the brain predicts the next step (e.g., step B). Once step B is received, the brain seeks the next step from a new key, "A->B," which incorporates the previous two steps. Essentially, the shortest key is employed initially to look for related sequences in long-term memory, returning a set of possible sequences as candidates. As additional information is extracted from long-term memory, the key continues to grow, and the candidates are further refined and filtered until the task is completed. Moreover, a representation in the key may be repeated, such as "A->B->A->B," and it is necessary to recall the previous steps accurately to correctly retrieve the sequence. This fundamental concept is further elaborated in Sect. 5.5, where NLP pattern matching is used as an illustration.

5.2 Conceptual Design

The chapter employs an n-bit integer data type for each input, which is more efficient than using floating-point representations. This approach helps save energy and space, while increasing throughput. The input data is processed sequentially, one datum at a time. Upon receiving a datum, it is stored in a shift register located in the working memory (i.e., the Hippocampus). The controller's counter increments by one, indicating the next shift register in the working memory. The input data is retained until the task is completed or the working memory is full.

Input data is processed and stored in sequential order, following the brain's method of processing and storing information in sequences. The Thalamus or controller manages the connection of each input data or shift register in working memory. When processing information in a sequence, the Thalamus only activates relevant information. Moreover, each sequence has a data count or counter, representing the significance of the information, which helps with counting and rehearsing of information. If a sequence occurs frequently, the data count will be high, indicating that the information is significant and should be rehearsed and stored in long-term memory. Figure 5.3 illustrates how the input is received and transferred to working memory, including a counter.

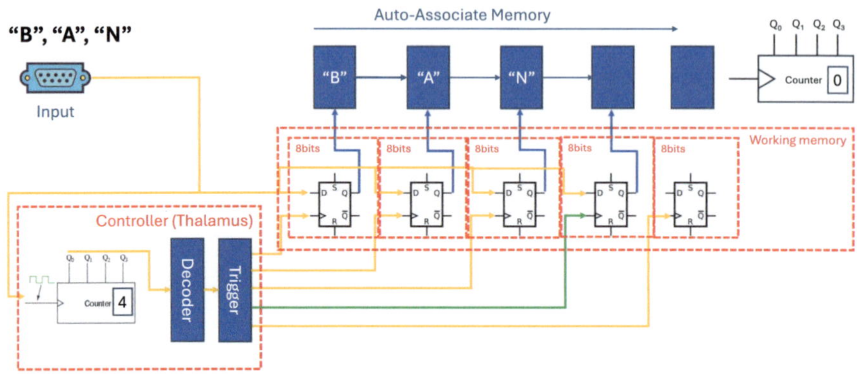

Fig. 5.3 How the new brain-inspired sequence learning memory works in the working memory and controller

If a sequence's data count exceeds the long-term memory threshold (which is typically set to three), it is promoted to long-term memory. If it fails to exceed the threshold, it remains in short-term memory. The data of a sequence will be eliminated over time, with the duration it's retained depending on the frequency of its appearances. However, when memory becomes full, the algorithm removes sequences with lower data counts first, as they are considered less relevant or significant. The sequence's data will be removed over time, how long it is kept depends on how many time it is seen. To form a sequence, input data is auto-associated with the previous input in working memory, and the algorithm predicts the next step based on past experience. If the prediction is incorrect or unexpected, the brain may switch to learning mode or correct the input after the task is complete. The working memory sequence is compared with all sequences in long-term memory. If a match is found, the discovered words are used to predict the next pattern. If the input is unpredictable, the information is retained in working memory until the task is completed. The complete architecture of the model can be found in Fig. 5.4.

After the task is completed, the algorithm compares the sequence data stored in working memory with all sequence data in long-term memory (Neocortex). If a match is found, the data count of the matched sequence is increased by one. If the number of different data is less than or equal to a noise threshold, the data in working memory is corrected using the sequence of data in long-term memory that has the smallest difference, since it is assumed to be noise. However, if the number of different data in the long-term memory exceeds the noise threshold, the algorithm enters the learning mode. The sequence in working memory is stored as a new sequence in short-term memory, and the data count is set to one. If the sequence already exists in short-term memory, the data count is incremented by one. The complete architecture of the model is depicted in Fig. 5.4.

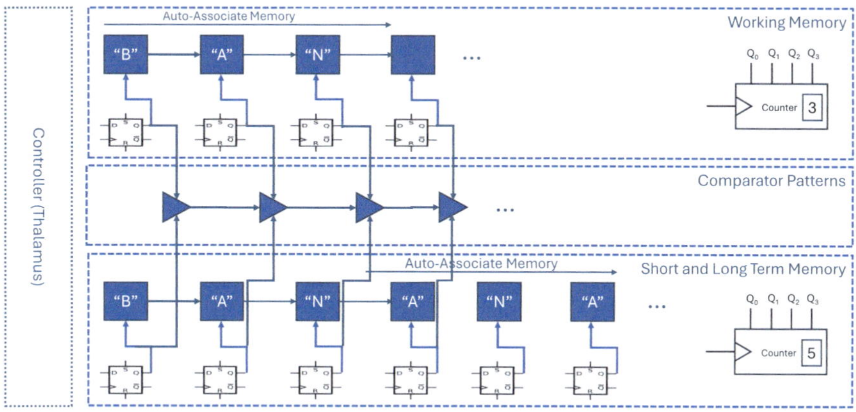

Fig. 5.4 The neocortex learning model architecture

Sequential memory works similarly to Random Access Memory (RAM) in terms of its structure, as shown in Fig. 5.5. Therefore, current technology's memory can be used for the algorithm. Additionally, the intention is to reuse generic hardware for cost efficiency and ease of adaptation.

5.3 Structure of Memory

SDRs are used as the memory structure throughout this chapter, supporting a hierarchical organization in which each representation can connect to the layer above or below.

An example of text encoding is shown in Fig. 5.6, where the first layer consists of multiple cortical columns, each represented by an active "1" bit in an SDR. Text elements can be encoded into any column, depending on the encoding method. This chapter introduces an encoding approach based on a hash function that maps not only individual characters to representations but also encodes connections between characters, mirroring how the brain encodes sequences. The second layer captures words and sentences, following the same principles of representation and connectivity as the first layer. All of this information is stored in an SDR, a large binary vector with only a small number of active bits, inspired by the brain's sparse coding strategy.

For example, in Fig. 5.6, the size of SDR is 10 bits. It contains a zero vector [0 0 0 0 0 0 0 0 0 0]. The representation of "I" is hashing "I", hash ("I") %10 = 3, the output is [0 0 0 1 0 0 0 0 0 0]. The representation of "EAT" is hashing "E", "A", "T", "EA", "AT" and "EAT"; the output is [0 0 1 1 0 1 0 0 0 1].

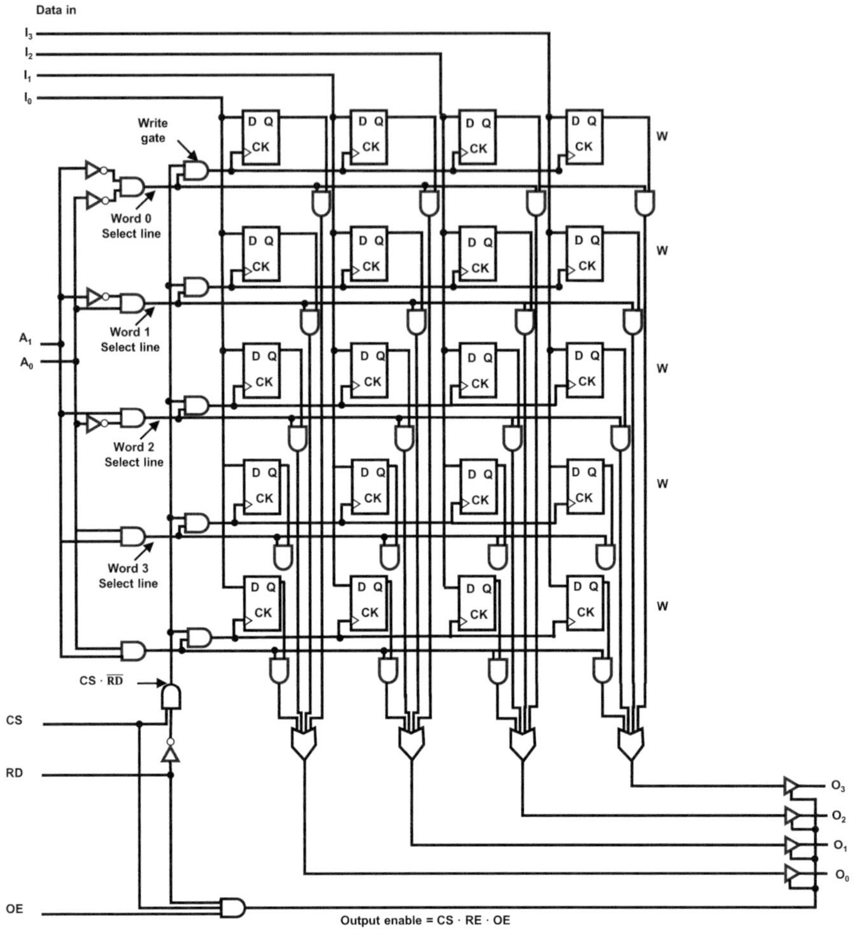

Fig. 5.5 Memory Structure

5.4 Fault Tolerance and Similarity

In this approach, determining whether an input is noise, already known, or new is critical. Instead of using weights to connect cells, bits or logical operations are used to process each representation. This reduces complexity and processing time, addressing a current issue in AI, while also supporting modern memory structures that store information in bits. For instance, characters are converted into bits, and connections are formed using a hash function, as illustrated in Fig. 5.7, layer 1. Each representation or word can be compared using logical operations. Additionally, sentence comparison in layer 2, or at the sentence level, can be achieved by concatenating each word and sentence representation from the lower layer. For instance, For instance, con-

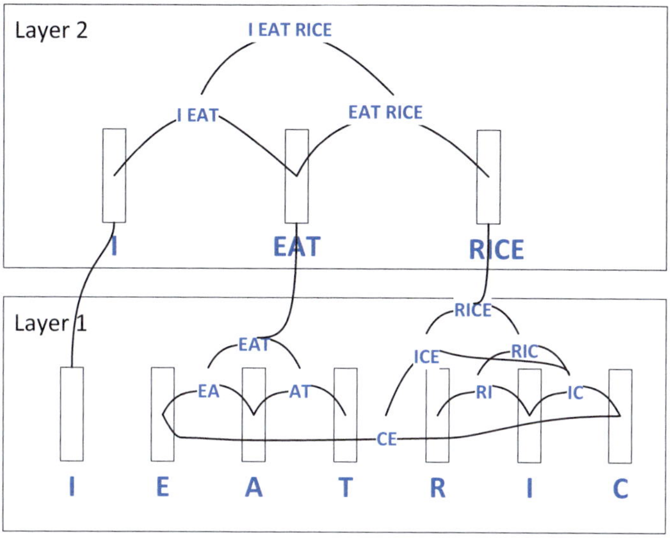

Fig. 5.6 Forming connections of each representation and sequence patterns in SDR

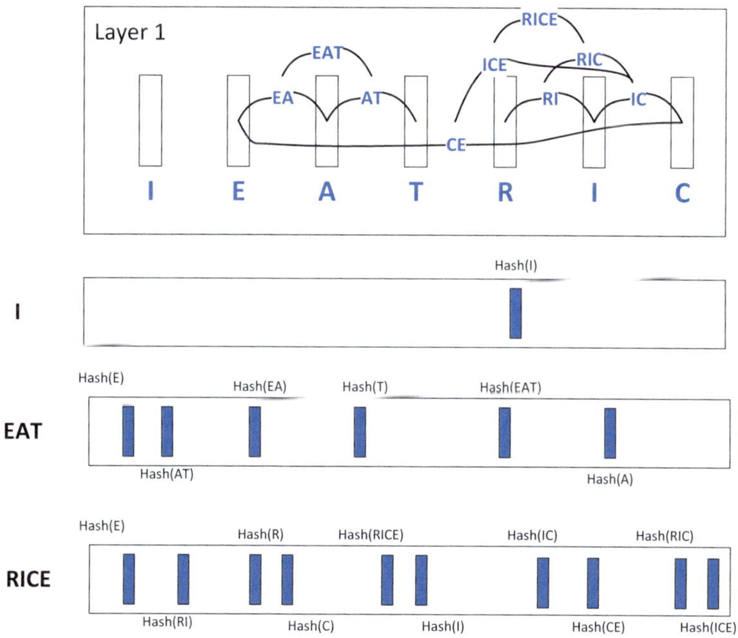

Fig. 5.7 Converting representations into bits by using a hash function

sider the SDR of the word "EAT" as [0 0 1 1 0 1 0 0 0 1]. When the misspelled word "EET" is received, it is converted to an SDR vector of [0 0 1 0 0 1 0 1 0 1], and the similarity can be computed using logical operations like AND or XOR, resulting in a similarity value called *diff*. In this example, the diff is 3 (AND). However, similarity comparison for "EET" is not only performed for the word "EAT," but for all word vectors in the dictionary, to find the highest number of diff (AND) or the lowest number (XOR). Since the operations are conducted at the bit level, the process is fast. At the sentence level, a similar approach is used by concatenating each word and sentence representation from the lower layer. For instance, the sentence "I EAT RICE" is concatenated by combining the three vectors, resulting in a size of 30 bits rather than 10.

5.5 NLP Pattern Matching

Sections 5.3 and 5.4 explain the usage and matching of structured SDRs. This summary provides an overview of the concept and gives an NLP example, illustrated in Fig. 5.8.

1. At the beginning of the process, the working memory is devoid of any information, while the neocortex contains previously encoded experiences in the form of sequences, as illustrated in Table 5.1. To represent each word sequence, it is converted into an SDR vector. For example, the word "BUT" is hashed by combining "B", "U", "T", "BU", "UT", and "BUT" to create an SDR vector of [0101010101010]. One important point to note is that the second counter will be reduced at each step, which could represent time or the step itself.
2. Once the brain perceives the first step, it searches for matching representations in the neocortex using the first hashed element ("*B*") as a key. SDR matching is used to determine which sequences in long-term memory match, with any short-term memory not yet stored excluded if it does not fit the representations. In this case, only the first two sequences in long-term memory match, resulting in the predicted elements of "*U*" and "*A*". These elements are connected to "*B*" through associative memory and correspond to two output lines that need to be activated, as they are predicted to be the next elements in the sequence. This activation of output lines is necessary for all potential continuations of the sequence in the $i+1^{th}$ step. However, in case no continuation is found, two possibilities arise:
   a. The brain has not yet identified the relevant representations, and therefore switches to a mode of learning.
   b. Upon encountering an error element, the brain employs fault tolerance to continue reading subsequent elements in an attempt to identify possible solutions.
3. Moving to the second step of the sequence, which is "*A*", the brain uses the hashes of both the first step ("*B*") and the second step ("*A*" and "*BA*") to search for a key. The search results in a match only with the hash pattern of "*BALL*". An essential

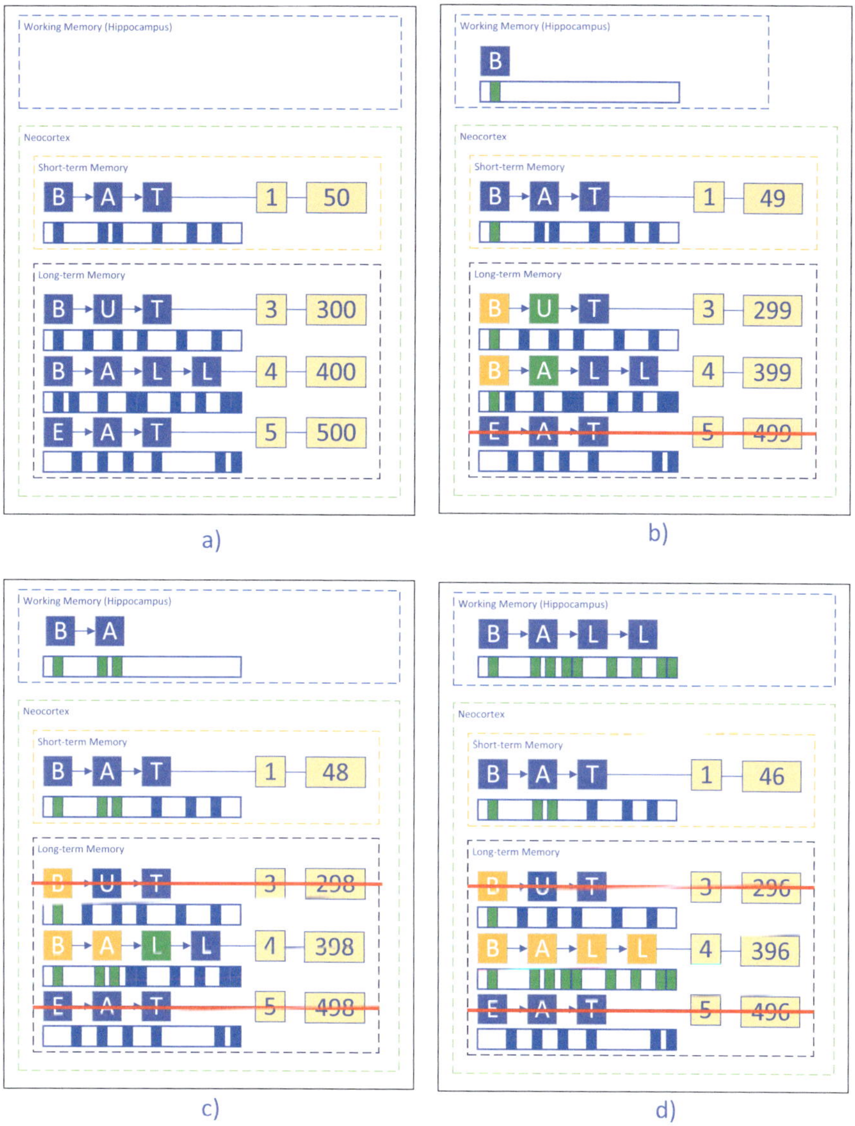

Fig. 5.8 Sequential learning in a new cognition-inspired sequence learning memory

concept to note here is that for each subsequent element in the sequence, the key space is extended by one, thereby increasing the number of keys used to address the content in the working memory. This means that the number of keys used to address the content should be increased in every step of the sequence reading process in the working memory.

4. The process continues until the task is complete, which happens in the fourth step when the element "*L* is perceived. Once the sequence is completed, the working memory will be released. This means that the SDR vectors associated with the sequence will no longer be stored in the working memory.

5.6 Brain-Inspired Method Versus Deep Learning

As indicated in Table 5.2, the Brain-Inspired model can work with smaller amounts of data and is easier to understand due to less complicated computations. It doesn't use backpropagation, which requires the entire dataset to be static and available during training and comes with a high computational cost. The model is also robust to noise and errors, and can operate both online and offline. It can be seen as more biologically accurate and requires less data and computational power.

Another key distinction lies in the architecture of deep learning versus Brain-Inspired methods. Deep learning relies on multiple interconnected layers of nodes, as illustrated in Fig. 3.15. In contrast, the Brain-Inspired approach uses a single-layer system that stores bit-level information in structures analogous to the neocortex or in mechanisms resembling short-term and long-term memory. This allows it to match new patterns with previously stored ones.

While implementing a multi-layer structure in Brain-Inspired methods is possible, it falls outside the scope of this book and is left as a direction for future research. The internal mechanism for storing memory and predicting the next sequence in Brain-Inspired models functions similarly to Long Short-Term Memory (LSTM) networks. Like LSTM, the Brain-Inspired approach maintains an internal state to support sequence prediction over time.

Table 5.2 Brain-Inspired versus deep learning

Attributes	Brain-Inspired	Deep learning
Data requirements	Small data	A lot of data
Interpretability	Understandable	Black boxes
Hardware requirements	Less computation	High computation
Learning	SDR (bits)	Weight and backpropagation
Noise	Robust	Sensitive
Adaptability	Online/Offline	Offline

Another contrast is that Brain-Inspired methods make use of Sparse Distributed Representations (SDRs) to identify similarities and classify the respective data class, bypassing the necessity for weight adjustment. This deviates from deep learning, which learns by modifying weights to reduce error. However, a notable benefit of Brain-Inspired methods is their capability to function online, thereby avoiding the necessity to freeze weights as in deep learning.

5.7 Hardware Simulation

The objective of developing a new method is to maintain simplicity while demonstrating the proof of concept through hardware simulation. The implementation of this method is achieved through the use of generic hardware to ensure cost-effectiveness and ease of implementation. MATLAB Simulink software is utilized in developing the new method, taking advantage of the HDL module to design AI hardware for FPGA and ASICs suitable for the algorithm, which requires real-time interaction with the environment. The software runs on an ASUS TUF A15 laptop equipped with an AMD Ryzen 7 5800H, 8 CPU cores, 16 threads, 32GB of RAM, and GPU RTX3060 6GB. The implementation is composed of five components, inspired by the brain, as shown in Fig. 5.9.

The system consists of five main components: input (sensory), output (motor), working memory (hippocampus), decision (Thalamus) and memory (Neocortex). The sensory or input component is connected to a communication port and is always prepared to receive input characters. Once it receives input, the module sends it to the thalamus for processing. The output or motor component is responsible for delivering the output from the algorithm to the user. It is linked to a communication port and employs the thalamus as a medium of communication to relay data to the user. The

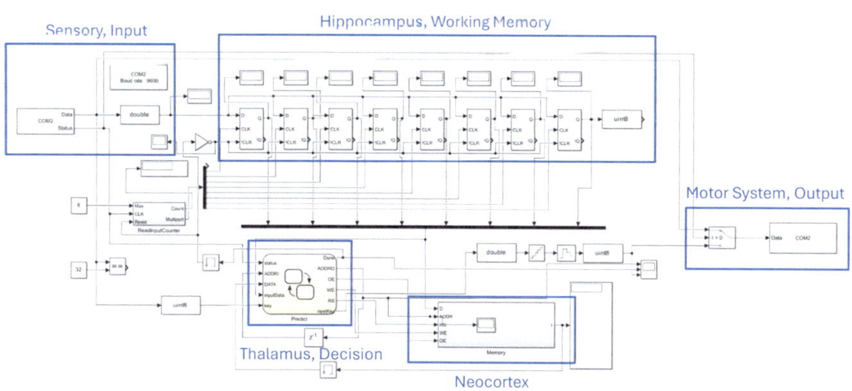

Fig. 5.9 Overall hardware architecture of the cognition-inspired sequence learning memory on MATLAB Simulink

working memory or hippocampus is used to retain information temporarily until a task is completed. Each input is processed by the thalamus or controller before being transferred to working memory. The data is stored in a sequential order to maintain its context.

The neocortex is responsible for both long and short-term memory and plays a crucial role in the learning process. Its primary function is to store information from the working memory once a task is completed. Upon receiving a sequence, the word counter is incremented by one if the sequence already exists in memory. If it does not exist, the sequence is added to memory as a new sequence, and the word counter is set to one. The memory operates under a threshold of three, where if the word count surpasses this threshold, it is transferred to long-term memory. If it does not exceed the threshold, the sequence remains in short-term memory. It is noteworthy that the sequence can be forgotten over time if it is not reinforced. In this implementation, the threshold for transferring information from short-term to long-term memory is set to three.

The thalamus or controller is responsible for encoding and transmitting both encoded input and input to the working memory. This implementation employs a 2048-bit integer data type for input data, which is fed into the algorithm one by one and stored in a shift register or address in the working memory until the task is completed or the working memory becomes full. As the sequence is read, the addressing is increased. This process is demonstrated in Fig. 5.10. The controller compares the representations in the working memory with all sequences in the long-term memory. If a match is found, the next sequence in the long-term memory is utilized for prediction, and the output is sent to the output component. If no match is found, the output is "*", indicating "unknown." Once the task is completed, the controller automatically corrects any incorrect information that is near a long-term memory pattern. The algorithm compares the output with a score called the "*diff*" to evaluate the similarity of patterns. If the diff is lower than the noise threshold, the incorrect word is corrected without learning. However, if the diff exceeds the noise threshold, the algorithm enters learning mode, transfers the word to the neocortex

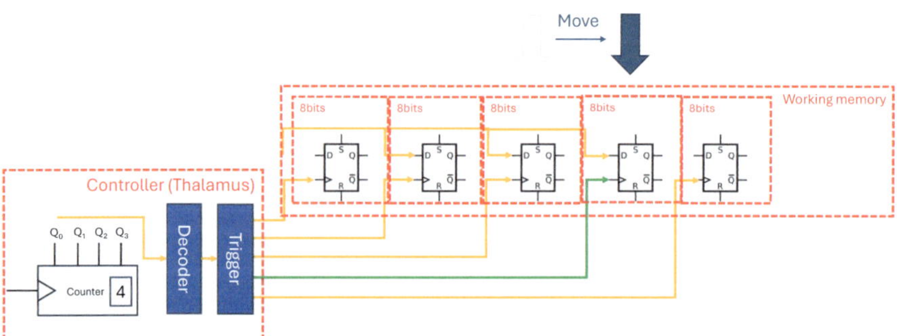

Fig. 5.10 Working memory in the new cognition-inspired sequence learning memory

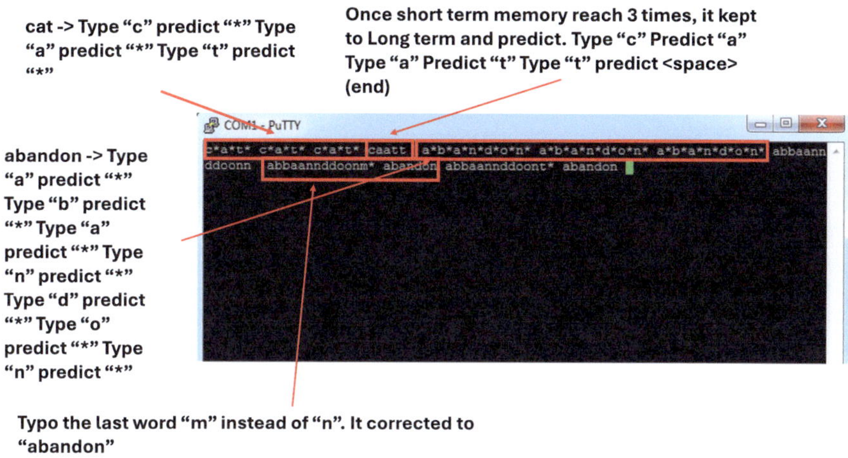

Fig. 5.11 Experiment to connect to the algorithm via comport for both input and output

to learn, and identifies it as a new word. The thalamus also relocates and forgets information from the working memory to the neocortex. Figure 5.4 displays a comparison of representations to determine the similarity score. The diff threshold can be calculated depending on the application or training data, such as the degrade5 dataset, by calculating the average difference between the ground words and input words of all training datasets. The average is 45, but it can be adjusted according to the application.

The experiment presented in Fig. 5.11 involves executing the Simulink MATLAB simulation on a laptop and linking it to the algorithm via a virtual comport. At the start of the test, both the neocortex and working memory are empty. The first input character received is "c", which is stored in the working memory. The algorithm compares the input's representations with those in long-term memory and generates a prediction of "*" or unknown as no information is available. The same prediction is made for the subsequent input characters "a" and "t". Upon pressing the <spacebar>, indicating task completion, the algorithm enters learning mode and learns the word "cat". The sequence "cat" is transferred from the working memory to the neocortex's short-term memory, and its word count is set to one. Then, the word "cat" is typed twice, causing its word count to increase to three and promoting it from short-term memory to long-term memory. When "c" is entered again, the algorithm discovers a match between working memory and long-term memory and predicts the next character, which is "a". Upon typing "a", the algorithm predicts the subsequent character, which is "t". This process is repeated for the word "abandon", but a typo is made in the last character, resulting in "abandom". The algorithm automatically corrects the mistake, and the correct output is "abandon" because the score is less than the noise threshold.

5.8 Conclusion

This chapter has introduced a brain-inspired sequence learning model that extends the concept of Sparse Distributed Representations (SDRs) to the temporal domain, focusing on how information is learned, retained, and predicted over time. Each section has been crafted to provide an integrated understanding of sequential cognition, starting with biological motivations and concluding with a practical hardware implementation.

The chapter began with Sect. 5.1, which introduced the cognitive process of sequence learning by explaining how the brain captures temporal patterns using working memory and reinforces them through repetition. This section was essential for framing the biological plausibility of the model and establishing the roles of short-term and long-term memory. Section 5.2 then translated these concepts into a functional architecture, detailing how input data is processed sequentially, stored temporarily, and promoted based on relevance. This laid the groundwork for an implementable system.

Section 5.3 introduced the SDR-based memory structure that encodes both representations and their associations across hierarchical levels, enabling the model to store complex sequences compactly. Building on this, Sect. 5.4 added robustness through logical bitwise operations that allow efficient noise filtering and similarity matching, which are critical for operating in real-world, imperfect environments. Section 5.5 demonstrated the application of these principles in natural language processing tasks, where the system must predict or learn sequences dynamically. The comparison in Sect. 5.6 contextualized the model against deep learning, showing how this approach offers interpretability, efficiency, and online learning capabilities without relying on backpropagation. Finally, Sect. 5.7 grounded the theory in practice through a hardware simulation, proving that the model can function in real time on generic, resource-constrained devices.

In this chapter established the theoretical foundation and proof of concept for a brain-inspired sequence learning memory, designed for simplicity, biological relevance, and ease of hardware implementation. Building on this, Chap. 6 applies the model to a practical task: spelling correction in natural language processing. This use case assesses the model's performance in real-world, noise-prone scenarios such as OCR errors and further demonstrates its practicality and efficiency in resource-constrained environments.

References

1. Fournier-Viger, P., Lin, C.W., U, R., Koh, Y.S., Thomas, R.: A survey of sequential pattern mining. Data Sci. Pattern Recogn. **1**(1), 54–77 (2017). https://doi.org/10.18063/dspr.v1i1.229
2. Zaki, M.J.: SPADE: an efficient algorithm for mining frequent sequences. Mach. Learn. **42**(1), 31–60 (2001). https://doi.org/10.1023/A:1007652502315

Chapter 6
Spelling Check Problem

Building on the sequence learning framework and hardware simulation introduced in Chap. 5, in this chapter shifts the focus from theoretical exploration to practical application. While Chap. 5 established how sequences can be learned and predicted using a simplified brain-like mechanism, it is crucial to assess the real-world viability of the model. In this chapter fulfills this need by applying the model to a highly relevant problem in Natural Language Processing (NLP): automatic spelling correction under noisy conditions, such as OCR-generated errors.

Spelling correction is a critical component in modern text processing pipelines. Whether in document digitization, search engines, or medical record systems, misspellings can lead to misinterpretation and degraded system performance. This chapter demonstrates how the proposed learning model can tackle this issue by offering fast, accurate predictions without extensive training, large datasets, or specialized hardware.

In this chapter begins by contextualizing the need for real-time spelling correction in NLP tasks, especially in environments where errors from OCR or user input can hinder system effectiveness (Sect. 6.1). It emphasizes the value of automatic correction mechanisms that reduce reliance on human intervention and heavy computational models. Section 6.2 reviews conventional spelling correction techniques, including dictionary-based, edit-distance, and deep learning approaches, highlighting their trade-offs in speed, accuracy, and resource requirements.

Section 6.3 describes how the brain-inspired memory model processes character inputs using working and long-term memory, enabling prediction and correction through sequence matching. It details mechanisms for handling noise, learning new sequences, and managing memory thresholds. Subsections introduce enhancements for scalability, such as parallel processing and vector merging, to support large datasets efficiently.

Section 6.4 evaluates the system on small and large datasets using OCR benchmarks. It compares the brain-inspired model against traditional and neural methods like pyspellchecker and ElmoSCLSTM, showing strong performance in both speed

T. Soisoonthorn and H. Unger, *A Brain-Inspired Approach to Natural Language Processing*, Studies in Computational Intelligence 1225,
https://doi.org/10.1007/978-3-032-00014-9_6

and accuracy, especially under noisy conditions. Sentence-level correction using n-gram context further improves results. Lastly, Sect. 6.5 concludes by highlighting the model's strengths in online learning, noise robustness, and low-resource deployment, demonstrating its practical viability for real-world spelling correction applications.

6.1 Introduction

In Natural Language Processing (NLP), researchers have studied ways to find and fix spelling mistakes for a long time. Spelling mistakes can make text harder to understand, so people use spelling checks to correct them and make the text easier to read. Many NLP applications need to correct mistakes before they can work properly. Nowadays, people use spelling checks in many different areas of life. For example, some programs can highlight incorrect words and suggest corrections to help people work faster and more conveniently, especially on mobile devices. Search engines [1] can also suggest corrections if they detect an error in a search query, and office programs can check for grammar errors and give suggestions to help users correct them.

In certain instances, interactive correction requires human interaction, leading to a delay in the process. As such, an automatic spelling correction system is proposed. For instance, in the context of medical records that are integral to diagnostic and treatment procedures, any error can have a significant impact on patients, medical research, and organizational workflows. Additionally, the digitization of documents, an effort aimed at reducing paper usage and storing information in databases, employs optical character recognition (OCR) to convert images into text. Spell-checking becomes a crucial aspect of the post-processing process because OCR alone cannot ensure flawless results. The key distinction between automatic and interactive spelling correction systems lies in the immediacy of the response. The latter requires immediate system response without the delay associated with human interaction.

6.2 Related Works

The first type of error that a spelling check can catch is called a non-word error. This happens when a word is spelled wrong and isn't in the dictionary. To fix this, the spelling check can use a method like Levenshtein Distance (LD) [2] or Damerau-Levenshtein Distance (DLD) [3] to compare the wrong word with words in the dictionary. LD uses edit, substitution, deletion, and insertion operations, but DLD is the same as LD, except it includes transposition. They both work by comparing the wrong word with all the words in the dictionary. This takes time, especially if there are a lot of words in the dictionary like in the Oxford dictionary, which has over 273,000 words. To speed up the search, researchers use techniques like a search tree [4] or hash table [5] based on Approximate string matching [6], which finds words that are close to the misspelled word rather than an exact match. Another method is brute force [7, 8], which tries all possible edits to a word to find the right one. This approach can generate many possible words, but it's time-consuming and limited to two edit distances. Another method is deep learning, which uses a machine-learning model like LSTM [9, 10] to learn from data and mistakes. This approach can be accurate but requires lots of training data and takes a long time to process without a GPU.

Sometimes, more than one word is often required. For example, the misspelled word "site" can also be "size" or "side," depending on the sentence. Thus, another real-world error type has been introduced. The word is misspelled, but it is still in the dictionary. A common approach to cope with this problem is to use its surrounding context. A common approach uses edit distance as an error model and n-gram for the context model. Deep learning is also used to cope with this problem by using the Sequence to Sequence Learning Model and integrating it with a transformer to improve the model [11]. This chapter also uses a deep learning model, Neuspell [12], with SC-LSTM plus ELMO (input), or Elmosclstm for comparison.

6.3 Algorithm

The spelling check problem is used to evaluate the brain-inspired concept and demonstrate its adaptation within the Brain-Inspired sequence learning memory. The proposed algorithm for solving the spelling check problem is illustrated in the flowchart (Fig. 6.1), Algorithm 1, and the steps that follow.

1. Receiving input into the working memory, the algorithm receives input in sequentially one by one and passes the input to Thalamus. The Thalamus moves the input to the working memory at the current register. Each input is received, and the counter is added by one. The current register will be moved as the value on the counter. This mechanism is shown in Fig. 6.2.
   a. At first, no data existed inside the working memory, and the counter was set to zero (0). Then, an input character "B" is received; the controller moves the input "B" to the first register as the counter is zero. After that, the counter is added by one.
   b. The new character "A" is received and moved to the second character as the counter is one and then added to two.
   c. After that, the new input "N" is obtained, the current register is used to keep the character, the counter continues to count, and so on.
2. Matching patterns between the working memory and the long-term memory, every time a new input is received, the algorithm finds the following pattern automatically to predict the next character. The controller checks for each sequence in the long-term memory to compare in comparators. The mechanism is shown in Fig. 6.3.

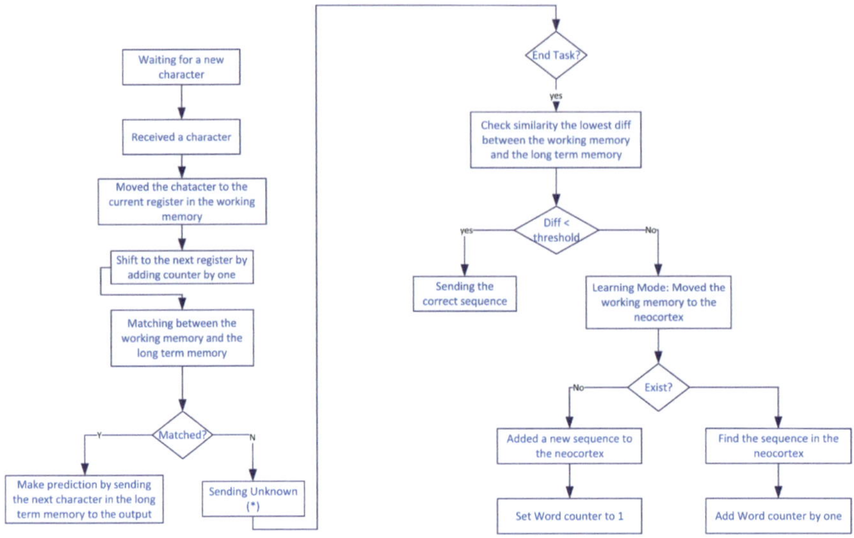

Fig. 6.1 Spelling check flow on the brain-inspired sequence learning memory

a. A new character "B" is received, and the controller will find matching in the neocortex as the key "B" by checking each sequence in the long-term memory. It found that the third row does not match with "B" as it starts with "C." It also found the second row as the word count below the threshold (3), which is not the long-term memory. Thus only the first and last row match, the next character that is auto-associated with "B" is "I" and "A" (Green). Thus, the output will be "I" and "A."
b. Once the input character "A" is received, it is auto-associated with the previous input "B" to create a sequence key "B->A" and find matches in the neocortex. Only the first-row match. The next character that is auto-associated with "B->A" is "Y," which is the current output.

3. If the sequence is in the working memory and the long-term memory match, it will predict and send the next character in the long-term memory to the output. Otherwise, it sends unknown or "*."
4. When the task is finished or the input contains a delimiter such as a <space bar>, and the sequence in the working memory does not match any in the long-term memory, the algorithm calculates the similarity score, or $diff$. If the $diff$ is not higher than a predefined noise threshold, the algorithm concludes that the input sequence is noise and outputs the sequence or word in the long-term memory with the lowest number of different characters. However, if the diff is higher than the noise threshold, the algorithm enters learning mode and identifies the input sequence as a new word that needs to be learned. The controller then moves the sequence from the working memory to the short-term or long-term memory,

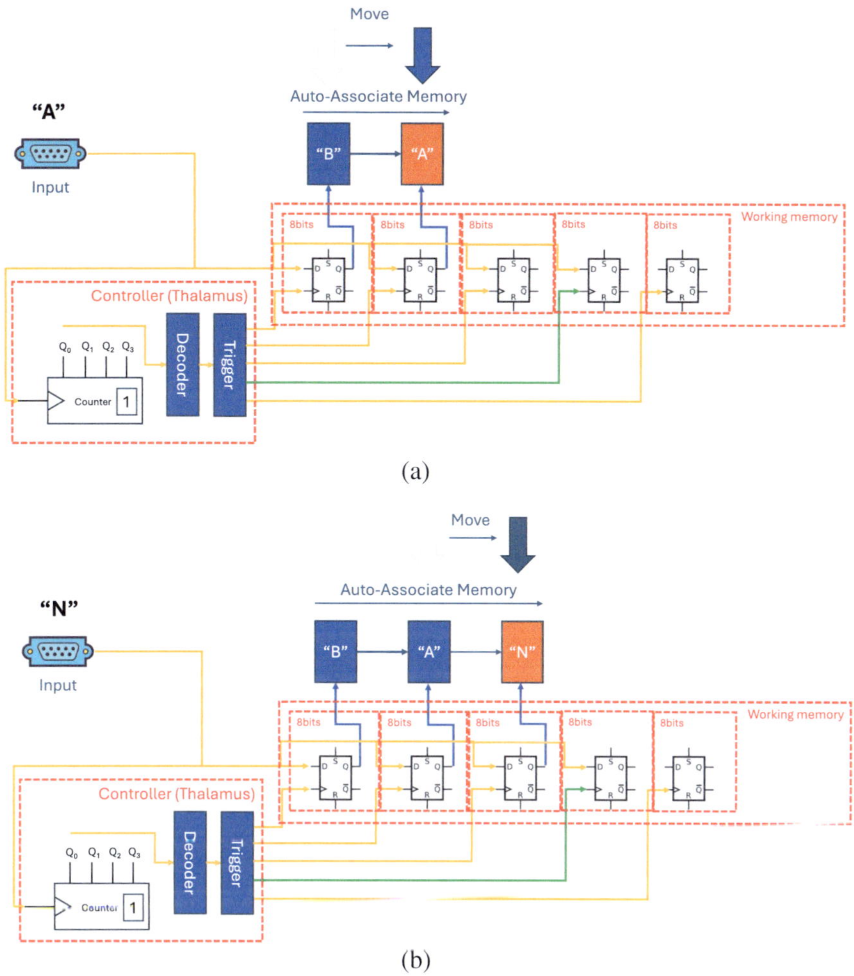

Fig. 6.2 Receiving input into the working memory

depending on the word count. If the sequence already exists in the memory, the word counter is incremented by one. Otherwise, the word counter is set to one (1).

6.3.1 *Moving From the Short-Term to Long-Term Memory*

The threshold for transferring information from short-term memory to long-term memory varies depending on the specific application. For instance, in offline learning for spelling check, the threshold can be set to 1 since the training data should consist

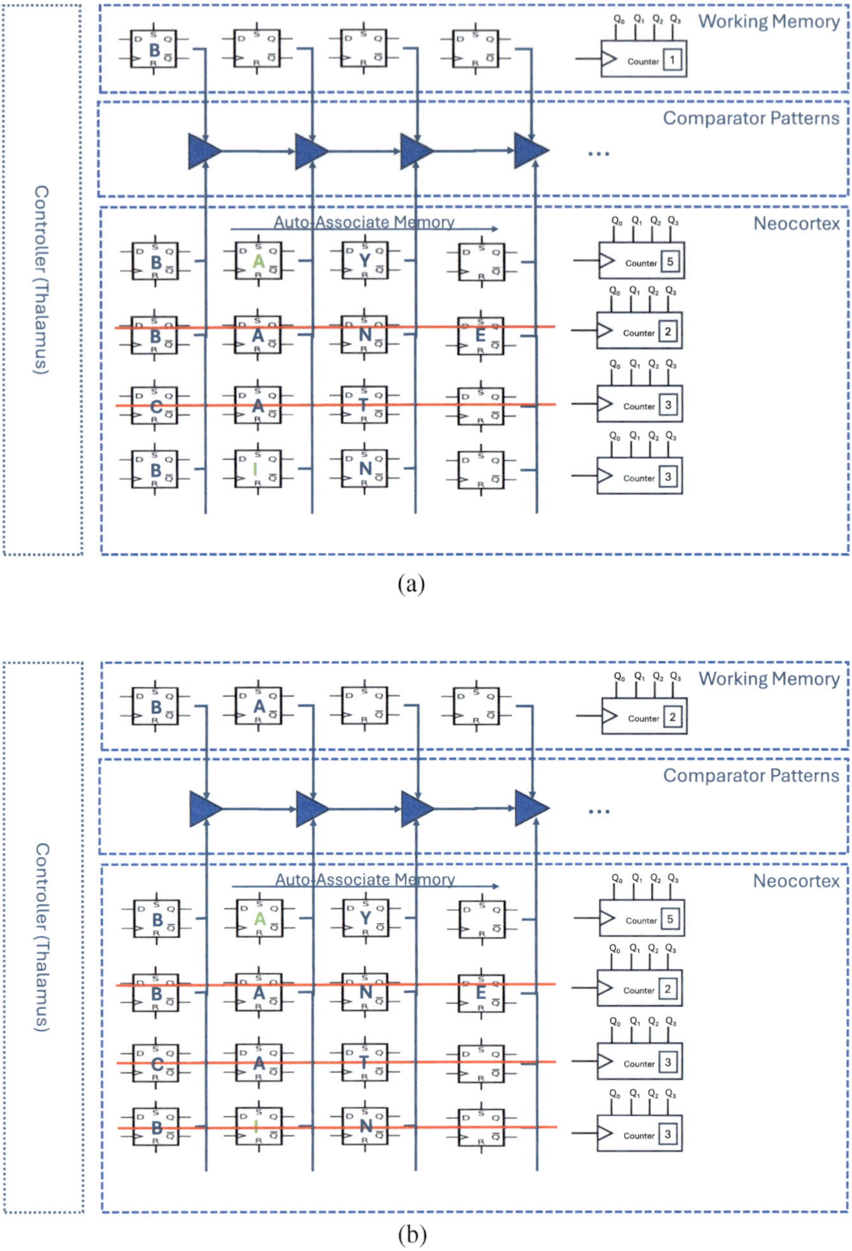

Fig. 6.3 Matching patterns between the working memory and the long-term memory

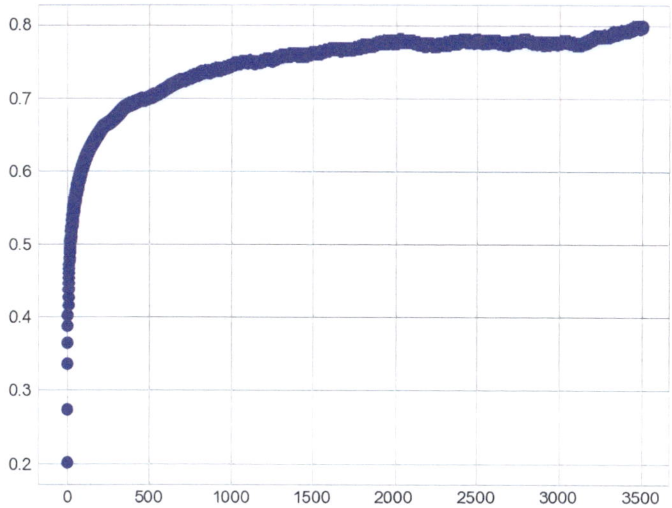

Fig. 6.4 The number of occurring words and the average percentage of correct learning from TREC-5 Degrade 5%

of correct words. In contrast, in hardware experiments in Sect. 5.6, the threshold is often set to 3 due to the intuitive belief that humans can remember something after encountering it three times. However, when it comes to online learning for OCR spelling check, the system's ability to learn from OCR text is called into question. OCR text may contain both correct and incorrect words, and the OCR system is expected to produce correct words more often than incorrect ones. For example, the word "eat" may appear 5 times while the word "ett" appears 2 times, making it reasonable to set the threshold to 3 when encountering the word "eat" in order to learn it correctly. The number of correctly occurring words and the average percentage of correct learning are shown in Fig. 6.4, and a threshold of 1000 can achieve 74.45% correct learning. However, some words may have more incorrect words produced by the OCR system than correct ones, as seen in the TREC-5 degrade5 data set where incorrect words account for around 20% of the total. Therefore, for OCR applications, it is advisable to learn only from correct words or sentences and set the threshold to 1 accordingly.

In some applications, such as correction of search queries, it is assumed that users will type more correct words than incorrect ones. Therefore, the system can learn from the input of search queries. For example, the qSpell [13] data set includes 11,134 queries randomly sampled from the publicly available AOL and 2009 Million Query Track. Figure 6.5 illustrates the number of correct words occurring and the average percentage of correct prediction of search queries. The results indicate that setting the threshold to 10 can achieve 98.09% correction learning, with only 109 words that were never corrected correctly. However, it should be noted that this data set is of low volume. With search engine characteristics and a larger data set, the number of correct words is expected to be higher than the number of incorrect words, which should improve the correction learning rate.

Algorithm 1 The Brain-inspired Sequence Learning Memory for Spelling Check

```
working_address ← 0
while not Program Exit do
    received_char ← read_char()                              ▷ waiting for a new character
    save_workingmemory(received_char, working_address)
    working_address ← working_address + 1
    neocortex_sequence_id ← 0
    prediction_character ← NULL
    min_diff ← MAXINT
    lowest_sequence_id ← NULL
    unexpected ← true                                        ▷ get word from working memory
    working_sequence ← get_workingmemory_autoassociate(working_address)
    while neocortex_sequence_id < MAX_NEOCORTEX_ROW do
        count ← get_neocortex_counter(neocortex_sequence_id)
        if count ≥ 3 then                                    ▷ only long term memory counter > 3
            if received_char = <space bar> then    ▷ end of the task    ▷ get word from long-term memory
                longterm_sequence ← get_longtermmemory_autoassociate(neocortex_sequence_id)
                diff ← similarity(working_sequence, longterm_sequence)
                if diff < min_diff then
                    min_diff ← diff
                    lowest_sequence_id ← neocortex_sequence_id
                end if
                if diff = 0 then
                    unexpected ← false
                end if
            else                        ▷ get word from long-term memory, limited by working_address
                longterm_sequence ← get_longtermmemory(neocortex_sequence_id, working_address)
                if working_sequence = longterm_sequence then
                    unexpected ← false
                    predict_character ← predict_longtermmemory(neocortex_sequence_id, working_address)
                    print(predict_character)
                    break
                end if
            end if
        end if
        neocortex_sequence_id ← neocortex_sequence_id + 1
    end while
    if unexpected = true then
        print(*)
    end if
    if received_char = <space bar> then
        if min_diff < threshold then                         ▷ noise then correct it
            longterm_sequence ← get_longtermmemory_autoassociate(lowest_sequence_id)
            print(longterm_sequence)
        else                                                 ▷ new then learn it
            save_neocortex(working_sequence)
        end if
        working_address ← 0
    end if
end while
```

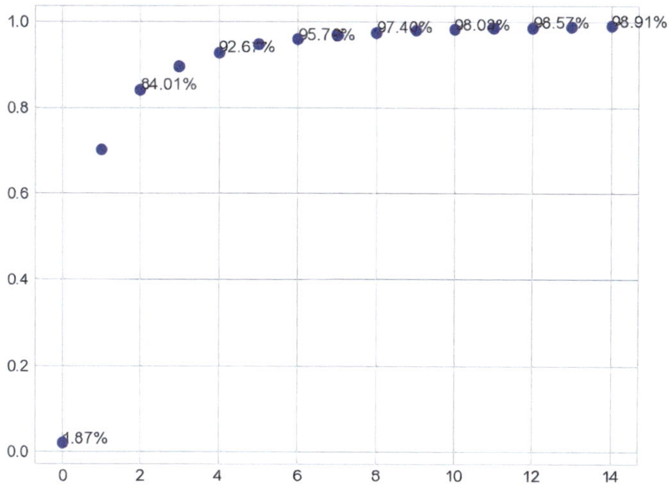

Fig. 6.5 The number of occurring words and the average percentage of correct learning from qSpell search queries

6.3.2 *Big Data Set*

While performing bit operations is fast, its speed decreases as the number of vectors grows linearly. For instance, an experiment showed that processing 7 million vectors can take up to 1 s. To cope with this issue, this chapter suggests two options that can be used in the evaluation process. The first option is to use parallel processing since modern processors come with multiple cores and threads. Parallel processing can significantly speed up the computation time by distributing the workload among the available cores and threads. The second option is to use merge or union vectors as defined in Sect. 4.6. This approach involves combining similar vectors to reduce the number of vectors that need to be processed. By doing so, the number of bit operations required to compute the similarity between vectors can be reduced, leading to faster computation times. Both of these options can be used in both offline and online scenarios, depending on the specific application and system requirements.

The process of merging in the context of vectors involves identifying the vector that is most similar to the target vector and aligning it with it. The similarity or distance between two vectors is checked to determine whether they can be merged into a single vector. To avoid a vector from becoming too dense or losing its uniqueness, two parameters are evaluated during the merging process. The first parameter is the sparsity of a vector, which is the percentage of active bits in the vector. The second parameter is the distance threshold, which is determined by plotting a histogram of the distance between all vectors, as shown in Fig. 6.6, and selecting a percentile value. For instance, if the percentile is set to 90, the distance threshold would be 1.975. Once merging is complete, sample words can be extracted for a vector, as shown in Table 6.1.

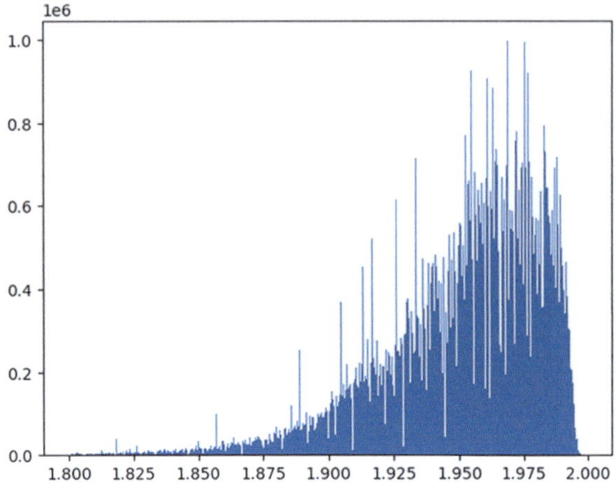

Fig. 6.6 Plotting histogram of all distance vectors to find a suitable percentile

Table 6.1 Merging vector example

Vector	Merged words
1	(ACF), (ACF), (A-F), (ACK)
2	"bug,", "bug", "bag,", "beg"
3	"baud", "bad", "bad", "baud"
4	"scab", "scab,", "scows", "scow", "scans"

6.4 Evaluation

The effectiveness and efficiency of the system are evaluated using two metrics: accuracy and average processing time. The evaluation is carried out on an ASUS TUF A15 laptop with an AMD Ryzen 7 5800H processor, 8 CPU cores, 16 threads, 32GB of RAM, and a GPU RTX3060 6GB. The standard TREC-5 Confusion Track [14] is used to evaluate the OCR spelling correction system. Each experiment is repeated five times, and the average value is calculated.

This dataset comprises two corrupted versions of 55,600 documents created by applying OCR to page images. The first version has an error rate of approximately 5% (degrade5), and the second is a down sample with an error rate of about 20% (degrade20). After cleaning, the dataset contains 3,532,743 lines, 33,255,482 words, and 701,217 unique words.

Table 6.2 compares the Brain-Inspired method with other methods such as Neuspell (Semantically Conditioned LSTM plus Elmo Transformer, SC-LSTM plus ELMO at input; Elmosclstm), LSTM, pyspellchecker, and Levenshtein Distance. Each method has different characteristics, as depicted in the table.

Similar to the Brain-Inspired method, pyspellchecker learns from dictionaries, but it uses the new vocabulary provided by the TREC-5 dataset instead of a built-in dictionary. On the other hand, LSTM and SC-LSTM plus ELMO require learning an error model for both correct and incorrect words. This can be divided into two

Table 6.2 Different types for each method

Method	Training correct words	Training incorrect words	Learning
Levenshtein distance	Yes	No	Offline
Pyspellchecker	Yes	No	Offline
LSTM	Yes	Yes	Offline
SC-LSTM plus ELMO (at input)	Yes	Yes	Offline
The brain-inspired	Yes	No	Offline/Online

areas of measurement: how the method manages incorrect patterns if the words are known, and how it deals with words that are not known. Dictionary-based systems will learn correct words from the training set and then test incorrect words within that same set to study how they handle incorrect patterns. Therefore, the training set itself will also be evaluated. However, this test cannot be applied to the error model-based methods, LSTM, and SC-LSTM plus ELMO, which learn from both correct and incorrect words in the training data. After learning from the training set, all methods will be tested on the testing set to evaluate how they handle unknown words. Furthermore, Elmosclstm and LSTM use a GPU for training and running, while LD, pyspellchecker, and the Brain-Inspired method use a CPU. The evaluation metrics used for comparison are accuracy and average processing time.

It's important to note that HTM isn't included in the evaluation. The architecture of the Brain-Inspired method significantly differs from that of HTM. Additionally, HTM hasn't previously been used for spell checking or word segmentation use cases. Furthermore, the state-of-the-art model is ElmoSCLSTM, which we have chosen for evaluation. Moreover, ElmoSCLSTM is the best model; it's a Transformer model based on NeuSpell, providing the highest accuracy.

There are three types of accuracy measurements used in this experiment. The first is Accuracy, which involves selecting the best score of words with only one candidate word. The second is AccuracyMax, which involves selecting multiple words with the highest score if they have the same score. Finally, AccuracyRange involves selecting a certain number of words (configured to 30 words) with the highest scores, which can also be checked again using Levenshtein Distance. For example, given an input word "W0rd" and candidate words with scores of "Word" (score 4), "Ward" (score 4), "W0ad" (score 3), and "Wo9a" (score 1), "Word" and "Ward" have the same score. Accuracy would select only one, "Word," but AccuracyMax would select both "Word" and "Ward." AccuracyRange would select "Word," "Ward," and "W0ad" if the configuration is set to three words.

To evaluate the performance of the Brain-Inspired method, two data sets were used: small and big data sets based on the degrade5 data set. The small data set consisted of 100,000 words for training and testing, with 12,961 unique words. On the other hand, the big data set used 80% of the degrade5 data set for training and 20% for testing. The purpose of using both data sets was to evaluate the effectiveness of the method in correcting a limited vocabulary, such as search queries, which can be achieved with a small data set, and a larger vocabulary, such as OCR correction, which may contain specific patterns, item numbers, and page numbers. The latter may require correction of words that are not in a dictionary, resulting in a larger set of words to be corrected.

6.4.1 Small Data Set

Table 6.3 compares the word-level performance of each method. The Brain-Inspired method shows a much faster response time compared to LD, pyspellchecker, and LSTM, approximately 300 times faster, and 14 times faster than Elmosclstm. This is due to the efficiency of bit operations, where the new method creates 12,961 SDR vectors with a 2048-bit size. In terms of accuracy, the Brain-Inspired method achieves an outstanding result of 91.62% with the training data, indicating that it can achieve a good result if it learns the correct words. However, if the correct words are not learned, it will only achieve 75.02%, which is similar to other methods. In comparison, the pretrained Elmosclstm model has an accuracy of only 54.18%, as it is a pretrained model and does not contain the characteristics of the OCR data set. However, it achieved a better result after training, reaching 82.18%.

Word-level correction is unable to fix real-word errors, as they require sentence-level context to be handled. In order to compare Elmosclstm and the Brain-Inspired method for sentence-level correction, the author analyzed their performance and presented the results in Table 6.4. The Brain-Inspired method uses two different

Table 6.3 Word-Level performance on a small data set

Method	Data	Accuracy	AccuracyMax	AccuracyRange	Time (s/word)
Levenshtein distance	Test	71.06%	75.11%	79.49%	0.390
Pyspellchecker	Train	70.75%			0.33
LSTM trained	Train	73.77%			0.328
LSTM trained	Test	71.94%			0.326
Elmosclstm pretrained	Test	54.18%			0.014
Elmosclstm trained	Train	82.18%			0.014
Elmosclstm trained	Test	75.29%			0.014
Brain-inspired 2048	Train	84.92%	91.06%	91.62%	0.001
Brain-inspired 2048	Test	71.07%	74.67%	75.02%	0.001

Table 6.4 Sentence-Level performance on a small data set

Method	Data	Accuracy	AccuracyMax	AccuracyRange	Time (s/word)
Elmosclstm trained	Train	81.97%			0.023
Elmosclstm trained	Test	77.65%			0.025
Brain-inspired 2-gram	Train	93.33%	96.98%	99.31%	0.021
Brain-inspired 2-gram	Test	56.55%	64.03%	64.03%	0.022
Brain-inspired 3-gram	Train	97.24%	98.64%	99.08%	0.024
Brain-inspired 3-gram	Test	46.83%	49.42%	55.93%	0.024

approaches, one based on 2-gram and one based on 3-gram, and concatenates two or three words, respectively, for every 2048 bits. The length of each vector depends on the number of concatenated words. Elmosclstm, on the other hand, provides a good result for the testing data, as it is already trained from pretraining. The Brain-Inspired method achieves the best accuracy of 99.31% when it knows the correct words. However, the accuracy of the test data is lower, with 64.03% for 2-gram and 55.93% for 3-gram, as some words were never learned during testing, and the dataset's vocabulary is small. This led to the identification of numerous unknown words. Sentence-level correction can provide a better result than word-level correction, as it can correct words based on their context in the sentence.

6.4.2 Big Data Set

Table 6.5 shows that the large dataset is divided into 80% for training and 20% for testing purposes. Table 6.6 displays the performance of each method at the word level on the big dataset. LD and pyspellchecker perform similarly to the small dataset, but LD's performance is slower at 13.14 s per word due to the large number of unique words (701,217) to search. LSTM was not evaluated on the larger dataset due to its considerable computational demands and time constraints. Furthermore,

Table 6.5 Big data set

	Line	All words	Unique words
Training 80%	2,780,211	26,604,373	701,217
Testing 20%	752,532	6,651,109	349,524

Table 6.6 Word-Level performance on a big data set

Method	Data	Accuracy	AccuracyMax	AccuracyRange	Time (s/word)
Levenshtein Distance	Train	80.83%	88.83%	94.83%	13.14
Pyspellchecker	Train	76.02%			0.30
Elmosclstm pretrained	Test	58.88%			0.017
Elmosclstm trained	Train	77.44%			0.019
Elmosclstm trained	Test	74.41%			0.020
brain-inspired 2048	Train	81.10%	87.38%	87.38%	0.05
brain-inspired 2048	Test	79.35%	87.03%	87.03%	0.05

considering that its performance does not exceed that of Elmosclstm, Elmosclstm can be used as the benchmark for comparison, acting as the state-of-the-art standard. Pretrained Elmosclstm has poor performance as it has not learned from the OCR pattern. However, after training, Elmosclstm achieves an accuracy of 74.41% for testing data and 77.44% for training data. Elmosclstm's time performance remains good as its architecture and parameters remain unchanged. This result is similar to the original paper, which achieves around 79.8%.

The Brain-Inspired method with 2048 bits and 701,217 SDR vectors performs better than Elmosclstm with an accuracy of approximately 5% for accuracy max and 10% for accuracy range. However, the performance time is roughly twice as slow as Elmosclstm. The result differs from the small dataset because the Brain-Inspired method has already learned enough correct words.

Overall, the performance of each method on the big data set indicates that the Brain-Inspired method and Elmosclstm are the most effective and efficient methods for word-level OCR correction.

The Elmosclstm model comprises approximately 209,906,438 parameters, taking up around 800MB of space. In contrast, the Brain-Inspired method utilizes 2048 bits and totals 701,217 vocabularies, amounting to about 171MB. This means the Brain-Inspired model's size is roughly 4.6 times smaller than that of the Elmosclstm model. The training time for Elmosclstm, which involves fine-tuning over 2 epochs, takes approximately 33 h. In contrast, configuring the Brain-Inspired method requires only about 154 s as it only needs to convert words into bits.

In order to enhance the system's performance, the author applied SDR vector merging as a preprocessing step, as multiprocessing can cause overhead that hinders processing time. The sparsity threads were set to 20% and the distance threshold was set to 0.2 based on the 90th percentile of the histogram plot. They evaluated the merging process on SDR vector sizes of 2048 and 4096 bits. After merging, the

Table 6.7 Word-Level performance on a big data set with merging

Bit size	Data	Merge vector size	Accuracy	AccuracyMax	AccuracyRange	Time (s/word)
4096	Train	39,811	80.27%	87.48%	87.86%	0.01
4096	Test	39,811	76.45%	83.85%	84.31%	0.01
2048	Train	78,074	78.57%	81.83%	82.02%	0.02
2048	Test	78,074	79.02%	82.22%	82.49%	0.02

Table 6.8 Sentence-Level performance on a big data set

Method	Data	Accuracy	AccuracyMax	AccuracyRange	Time (s/word)
Elmosclstm trained	Train	86.50%			0.015
Elmosclstm trained	Test	79.37%			0.017
Brain-inspired 4096	Train	100.00%	100.00%	100.00%	1.26
Brain-inspired 4096	Test	93.66%	93.74%	94.15%	1.95

vector sizes were reduced to 39,811 for 2048 bits and 78,074 for 4096 bits from the original size of 701,217.

Table 6.7 demonstrates the outcomes of merging SDR vectors on accuracy and processing time. The accuracy for 4096-bit sizes is close to that before merging, while the processing time is significantly reduced by a factor of 20 due to the reduction of the number of vectors. On the other hand, the 2048-bit size shows lower accuracy, as the increase in the size of the vectors can aid in reducing false positives. Consequently, for improving performance through merging SDR vectors, it is recommended to use the 4096-bit size.

Table 6.8 displays the sentence-level performance results on the big data set. Elmosclstm's performance is better than its performance at the word level, as it utilizes sentence-level information for prediction while maintaining a good processing time. The Brain-Inspired method exhibits better accuracy than Elmosclstm for both the training and testing data sets, particularly in the accuracy range metric, which yields 100.00% for the training data and 94.15% for the testing data.

Another experiment provides evidence that the Brain-Inspired learning method is highly noise-tolerant without requiring retraining of the error model. Even when tested on a degrade20 dataset, the Brain-Inspired method still achieves high accuracy using the same training data from degrade5 that shows a good result of 100% and 93.08% for the accuracy ranges of both training and testing data, respectively, as shown in Table 6.9. In comparison, Elmosclstm with retraining from degrade5 performs poorly on the degrade20 dataset, highlighting the advantage of the Brain-Inspired method in handling noise and errors. This robustness is essential in real-world applications where datasets can have varying levels of noise and errors.

Table 6.9 Sentence-Level Performance on A Big Data Set with A Degrade20 Data Set

Method	Data	Accuracy	AccuracyMax	AccuracyRange	Time (s/word)
Elmosclstm trained	Train	55.31%			0.016
Elmosclstm trained	Test	52.39%			0.017
Brain-inspired 4096	Train	100.00%	100.00%	100.00%	1.13
Brain-inspired 4096	Test	92.32%	92.64%	93.08%	1.25

Table 6.10 Sentence-Level performance on a big data set with parallel processing

Method	Data	Accuracy	AccuracyMax	AccuracyRange	Time (s/word)
Brain-inspired 4096	Train	85.10%	91.77%	93.24%	0.10
Brain-inspired 4096	Test	82.00%	83.12%	84.69%	0.13

Table 6.11 Sentence-Level performance on a big data set with merging

Method	Data	Accuracy	AccuracyMax	AccuracyRange	Time (s/word)
Brain-inspired 4096	Train	75.80%	85.95%	89.62%	0.36
Brain-inspired 4096	Test	73.73%	77.39%	82.03%	0.357

The Brain-Inspired method has a slow processing time of 1.26 s for training data and 1.95 s for testing data, which is not suitable for many applications. To address this issue, the method is parallelized using threading, which allows for easy implementation of parallel processing. Input is submitted to each thread, and the results are returned and combined in the main thread for score sorting. The results of parallel processing are presented in Table 6.10, with the number of threads set to 10.

Table 6.11 demonstrates that the data is partitioned into ten parts using parallel processing, which results in a significant reduction in processing time to approximately 0.1 s. However, there is a risk associated with this approach because the method calculates scores from all SDR vectors and sorts them using Levenshtein distance. In other words, it searches for word patterns that are similar to the input and then sorts them based on edit distance. As a result, candidates are chosen from only a portion of all word patterns by dividing the information patterns into each thread. This means that candidates with lower scores for checking SDR similarity may be included in the candidate list and ultimately selected because they provide the lowest LD distance, even though they may have a different pattern. Nonetheless, the results obtained are still superior to those of Elmosclstm.

Another method used in this study involves merging SDR vectors based on two parameters: sparsity and distance threshold. The vector merging process is sensitive to these parameters, with low sparsity or high distance threshold preventing vectors from being merged, while high sparsity and low distance threshold may lead to low accuracy. The merging process is applied to the word vectors obtained from the experiment in Table 6.8, with the resulting merged SDR vectors concatenated at the sentence level. For this experiment, sparsity is set to 0.15 (15%) and the distance threshold is 1.5. This reduces the number of vectors from 3,972,879 to 908,410, or approximately 4.37 times. Table 6.11 shows that the accuracy, accuracy max, and accuracy range of this method outperform Elmosclstm, even after reducing the number of SDR vectors. However, the processing time is not still slower than Elmosclstm, but still acceptable for most applications.

6.5 Conclusion

In this chapter, the Brain-Inspired model has demonstrated its superior time performance on small datasets due to its operational reliance on bits. The speed advantage is manifested not only during evaluation but also in training, which merely requires the encoding of data into bits. In contrast, the Elmosclstm model presents slower performance, attributed to its complex structure, a large number of parameters, and a necessity for sequential data processing. The experimental results also revealed that the Brain-Inspired model has a smaller size compared to Elmosclstm. This is due to the bit-oriented information storage approach of the Brain-Inspired model, while Elmosclstm requires a higher number of parameters for processing. However, it is important to note that the size of the Brain-Inspired model increases with data size, which contrasts with the fixed parameter quantity in Deep Learning models. To handle such growth, three mechanisms are proposed: parallel processing, a multi-layer hierarchical structure, and the use of Sparse Distributed Representations (SDR) unions. While this chapter encompasses parallel processing and SDR union usage, the exploration of the multi-layer hierarchical structure lies outside its scope. The Brain-Inspired model also outperforms Elmosclstm in terms of accuracy, particularly in noisy situations. This is a consequence of its hashing technique which stores the sequence information for each word, capturing not just the characters but their interrelations. Consequently, even when some information changes, the existing information in the SDR can still recognize the pattern. In contrast, Elmosclstm's accuracy is compromised when the error model from the application changes, as it fails to recognize the altered pattern. Enhancements to the accuracy of the Brain-Inspired model were also noted with the increase of n-grams, as the surrounding context enables the correction of real-word problem types from subsequent words.

In summary, the Brain-Inspired approach presented in this chapter offers a novel and efficient solution for spelling correction tasks. Its ability to work in both offline and online learning modes sets it apart from other methods that work only in offline modes. Additionally, the approach eliminates the need to learn from error words, which can be a challenging task, by relying solely on correct words and sentences.

The approach is also suitable for interactive spelling correction systems, as it provides feedback and predictions in real-time. Furthermore, it offers a new solution for small and big data sets, providing fast and accurate results if trained with the correct words and sentences.

One of the key strengths of the approach is its ability to run on commodity hardware, making it a cost-effective solution for spelling correction tasks. Lastly, the approach provides high noise tolerance, which is an important feature for tasks that involve noisy input. Overall, the contribution of this chapter offers a promising solution for improving spelling correction in various applications.

To further validate the generalizability and robustness of the brain-inspired model, Chap. 7 applies it to another challenging task: Thai word segmentation. This distinct use case builds on the core principles introduced earlier and demonstrates the model's adaptability across different languages and problem domains.

References

1. Martins, B., Silva, M.J.: Spelling correction for search engine queries. In: Soares, C., Machado, P., Costa, V. (eds.) Progress in Artificial Intelligence, Lecture Notes in Artificial Intelligence, vol. 3230, pp. 372–383. Springer, Berlin, Heidelberg (2004). https://doi.org/10.1007/978-3-540-30228-5_33
2. Levenshtein, V.I.: Binary codes capable of correcting deletions, insertions and reversals. Soviet Phys. Doklady **10**(8), 707–710 (1966)
3. Damerau, F.J.: A technique for computer detection and correction of spelling errors. Commun. ACM **7**(3), 171–176 (1964)
4. Shang, H., Merrett, T.H.: Tries for approximate string matching. IEEE Trans. Knowl. Data Eng. **8**(4), 540–547 (1996). https://doi.org/10.1109/69.536247
5. Mosavi Miangah, T.: FarsiSpell: a spell-checking system for Persian using a large monolingual corpus. Literary Linguist. Comput. **29**(1), 56–73 (2013). https://doi.org/10.1093/llc/fqt008
6. Ukkonen, E.: Algorithms for approximate string matching. Inf. Control **64**(1–3), 100–118 (1985). https://doi.org/10.1016/S0019-9958(85)80046-2
7. Norvig, P.: How to Write a Spelling Corrector (2023). https://norvig.com/spell-correct.html. Accessed 4 May 2025
8. Barrus, T.: pyspellchecker-Pure python spell checker based on work by Peter Norvig (2018). https://pypi.org/project/pyspellchecker. Accessed 4 May 2025
9. Kinaci, C.: Spelling correction using recurrent neural networks and character level N-gram. In: 2018 International Conference on Artificial Intelligence and Data Processing (IDAP), pp. 1–4. IEEE (2018). https://doi.org/10.1109/IDAP.2018.8620899
10. Sooraj, S., Kumar, M., Kp, S.: Deep learning-based spell checker for Malayalam language. J. Intell. Fuzzy Syst. **34**(3), 1427–1434 (2018). https://doi.org/10.3233/JIFS-169438
11. Sutskever, I., Vinyals, O., Le, Q.V.: Sequence to sequence learning with neural networks. In: Ghahramani, Z., Welling, M., Cortes, C., Lawrence, N.D., Weinberger, K.Q. (eds.) Advances in Neural Information Processing Systems 27 (NeurIPS 2014), pp. 3104–3112. MIT Press, Montreal (2014)
12. Jayanthi, S., Pruthi, D., Neubig, G.: NeuSpell: a neural spelling correction toolkit. In: Proceedings of the 2020 Conference on Empirical Methods in Natural Language Processing: System Demonstrations, pp. 158–164. Association for Computational Linguistics, Online (2020). https://doi.org/10.18653/v1/2020.emnlp-demos.21

13. Ganjisaffar, Y., Zilio, A., Javanmardi, S., Cetindil, I., Sikka, M., Katumalla, S.P., Khatib-Astaneh, N., Li, C., Lopes, C.: qSpell: spelling correction of web search queries using ranking models and iterative correction. In: Spelling Alteration for Web Search Workshop, Bellevue, WA (2011)
14. Kantor, P., Voorhees, E.: The TREC-5 confusion track: comparing retrieval methods for scanned text. Inf. Retrieval **2**(2–3), 165–176 (2000). https://doi.org/10.1023/A:1009902609570

Chapter 7
Thai Word Segmentation

After demonstrating the effectiveness of the Brain-Inspired sequence learning memory in spelling correction tasks in Chap. 6, in this chapter extends the evaluation of the Brain-Inspired memory model to Thai word segmentation. This task poses unique challenges due to the absence of explicit word boundaries. Building on the foundation from the previous chapter, this chapter explores whether the same principles can be effectively adapted to a new linguistic problem.

Through the introduction of two methods, THDICTSDR and THSDR, the chapter examines the model's ability to handle ambiguity and noise without relying on large datasets or complex architectures. Studying this chapter offers insight into the model's broader potential and encourages further reflection on its adaptability across diverse NLP tasks.

In this chapter begins by highlighting the challenge of Thai word segmentation, where the absence of clear word boundaries requires more adaptive solutions (Sect. 7.1). Section 7.2 reviews past methods such as rule-based approaches, dictionary-driven techniques, and deep learning models, noting their limitations in handling ambiguity and unknown words.

Section 7.3 introduces two brain-inspired methods, THDICTSDR and THSDR, which use Sparse Distributed Representations (SDRs) to encode words and their context. These models improve segmentation by matching patterns even under noise. Section 7.4 evaluates both methods on benchmark datasets, showing competitive accuracy and strong noise tolerance. Section 7.5 concludes that SDR-based models effectively adapt to Thai segmentation and offer a promising alternative to traditional and deep learning approaches.

T. Soisoonthorn and H. Unger, *A Brain-Inspired Approach to Natural Language Processing*, Studies in Computational Intelligence 1225,
https://doi.org/10.1007/978-3-032-00014-9_7

7.1 Introduction

The significance of Natural Language Processing (NLP) has been growing for numerous applications, including emotion detection, data extraction, text grouping, automated translation, voice recognition, and query responses. For languages like English, the process of word tokenization tends to be simple as it involves recognizing separator characters such as spaces, colons, commas, quotation marks, and full stops. Conversely, for languages without explicit segmentation, like Thai, Chinese, and Japanese, determining word borders poses a complex problem that necessitates specific algorithmic solutions.

The initial development of Thai word segmentation took place in 1981, which can essentially be categorized into three methodologies [1]: rule-oriented, lexicon-oriented, and learning-oriented techniques. Rule-oriented techniques involve the rigid implementation of rules. However, their scope in dealing with all potential cases is limited, and the creation of new rules might be necessary for unknown words. Lexicon-oriented methods rely on dictionaries to identify matches by analyzing character sequences, but their effectiveness is contingent upon the size of the dictionary, the strategy for dealing with unknown words, and the confusion that comes up when multiple segmentations are viable. To resolve such confusion, one technique is to choose the lengthiest word, although this isn't consistently reliable. An advanced strategy is to employ learning-oriented methods that take the word's context into account, thereby enhancing the precision of segmentation.

The learning-oriented methodology for segmenting Thai words involves clearly identifying word boundaries and applying machine learning techniques to build a predictive model. Notable strategies in this category include the Hidden Markov Model (HMM) [2], Conditional Random Fields (CRF) [3], and deep learning approaches such as Deepcut [4] and Attacut [5]. The strength of this methodology lies in its independence from dictionaries, allowing it to handle ambiguity and unknown words using statistical patterns. However, it requires a labeled training dataset, and its accuracy depends heavily on factors such as the domain of the training data, the size of the dataset, and the consistency of boundary annotation. Preparing such datasets can be both time-consuming and labor-intensive.

To address the challenges outlined above, this study introduces two innovative techniques. The first method, called THDICTSDR, is based on a dictionary and leverages Sparse Distributed Representations (SDRs) combined with n-grams. It proposes a novel encoder for Natural Language Processing (NLP), integrating it with n-gram structures. Results show that this brain-inspired approach can achieve a level of precision comparable to deep learning methods. The second method, called THSDR, uses SDRs directly, without relying on a dictionary, for word identification. This approach improves noise tolerance, making it especially effective in unfiltered data scenarios such as Thai OCR.

7.2 Related Works

The first rule-based [6] is created base on Thai grammar, and [7] improves the rules by using Thai spelling principles, and many rules are used in combination manners. The first dictionary-based [8] was proposed in 1986 using the dictionary combined with the longest matching technique by selecting the word with the longest length. If selecting cannot find the rest of the sentence, it will backtrack and find the next longest word. However, it will fail if the correct word is not the longest one or multiple unknown words are found. The maximal matching algorithm [9] was proposed to cope with the longest matching problem by finding all possible segmentation for a sentence and choosing one that contains the fewest words. Nevertheless, finding all possible words are a brute force method in a long sentence, and many candidates are generated. Therefore, the method does not guarantee the selection of the correct one and cannot determine the best one if candidates have the same number.

Another improvement of the dictionary-based is a Trie Structure that was proposed in 1991 [10]. Instead of keeping all words in a list, the method creates a structure like a Tree to reduce the storage size and find words by moving in a tree node for each character. Thus, it can find the word via Trie, which is faster than finding a word from a list. TLS-ART-MC [1] proposes a combination of Ranking Trie, Soundex, and two-pass segmentation. Ranking Trie is sorting character node that depends on their frequency. A higher frequency node will be closer to the root node, and the node will be found first. Thus, it makes the Trie structure smaller and performs faster. Soundex is used to cope with misspelling problems. The Soundex is designed for searching for an expected name with a different spelling. A word is converted to a code using rules; thus, the same code means the same word. Once the text passes segmentation from Trie and Soundex, the segment words will be combined using rules based on Thai grammar.

Another approach is learning-based by using the statistical technique, a Viterbi-based, [2] to employ statistical information derived from grammatical tags. The idea is to find the path that gives the maximum probability. Moreover, [11] and [12] propose using the word trigram with the part of speech. However, the method only captures its state and corresponding words, but the state might depend on something other than its following words or other adjacent words. In machine learning research, word segmentation is a classification problem that defines each character in the string as one of the binary classes, the beginning of a word labeled "B" and an intra-word character labeled “I”. These labeled characters and segmented words are trained to machine learning algorithms. Comparative research [13] is proposed to compare four learning-based algorithms that include Naïve Bayes (NB), Decision Tree, Support Vector Machine (SVM) and Conditional Random Field (CRF), and dictionary-based algorithms, the longest and maximal matching. The result shows that the dictionary-based algorithms perform better than NB, Decision Tree, and SVM. However, the best result is the CRF algorithm.

Instead of using segmentation words as a base, some research uses syllable or smaller group of characters that presents the performance better. [14] uses two processes, syllable segmentation, and syllable merging. Firstly, the research defines about 200 syllable patterns and trigram statistics for syllable segmentation. Then, merging by determining possible sequence word segmentation by dictionary and select the best word segmentation from the maximum collocation strength, which is the sum of co-occurrent within a word minus co-occurrent between words. [3] proposes using Minimum Text Units to extract the smallest units that constitute words and then using CRF to identify syllables and merge syllables by a set of rules. Another research proposes grouping Thai contiguous characters into inseparable units called Thai Character Clusters (TCCs) [15] that are defined by a set of rules. However, this research is provided for Information Retrieval to improve search accuracy.

The word segmentation state-of-the-art uses deep learning that can learn from data and error. A popular Thai word segmentation is Deepcut [4], which uses Convolutional neural networks (CNNs) that the accuracy from the experiment can achieve 96.57% and F1 at 96.34%. An improvement of Deepcut is Attacut [5] which uses syllable embedding as features together with character embedding. Even the accuracy and F1 of the research are not over Deepcut, but the processing time is at least 5.6 faster than Deepcut. [16] provides word segmentation and POS tagging by jointing models of both tasks to improve overall accuracy. The word boundaries are produced first and become an input for tagging. The input is character-level n-gram, and the next layer incorporates the n-gram features with their surrounding contexts by using bidirectional recurrent neural networks (RNNs). The accuracy results are above 90% on F1-Score.

7.3 Algorithm

7.3.1 Training

In the methodology proposed here, each word within the dictionary is kept in a HashMap structure, which associates the word with its corresponding SDR list. The foremost SDR in the list is allocated for encoding the word itself, while the rest of the SDRs in the list serve to encode the context around the word. The range of the surrounding context in Thai word segmentation is determined by a threshold value, typically set to 16 by default. This threshold value determines the number of surrounding words, including both suffix and prefix words, that are considered as part of the context for segmenting a particular word. Figure 7.1 graphically represents the storage of the words "perform" and "performance" in the map structure, together with their associated SDRs for the word itself and its surrounding context.

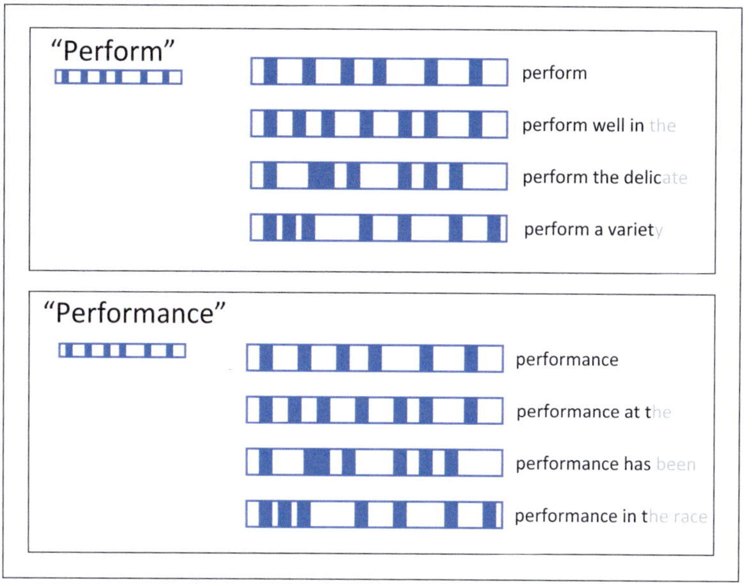

Fig. 7.1 The structure of words and SDRs

7.3.2 *Matching*

The THDICTSDR algorithm employs a blend of dictionary use and SDRs. It initiates the process by identifying a list of potential words in the text via the dictionary. If a single word is discovered, it's chosen. However, in the event multiple words are identified, the algorithm selects the appropriate word based on the degree of similarity between the SDRs of the surrounding context of each word and the SDR of the input text. For instance, as shown in Fig. 7.2, within the text "performanceatthemusic," the possible words are "perform" and "performance". The algorithm then scrutinizes the SDRs of the surrounding context of each word, such as "perform well in" and "performance at t". The highest SDR value similarity is between the SDR of "performanceatthemusic" and "performance at t", indicating that "performance" is the apt selection. The algorithm utilizes a threshold value to dictate the quantity of words to incorporate into the SDRs of the surrounding context, which could be suffix words, prefix words, or both.

THSDR adopts a method akin to THDICTSDR, but rather than scouring for a list of potential words, it employs SDRs to detect words. The application of SDRs offers a key advantage as they are tolerant to faults, enabling the algorithm to recognize words even when certain characters are absent or altered. After identifying candidate words, the algorithm chooses the most suitable match based on their surrounding context. This surrounding context is also trained and stored within SDRs, as demonstrated in the previous example.

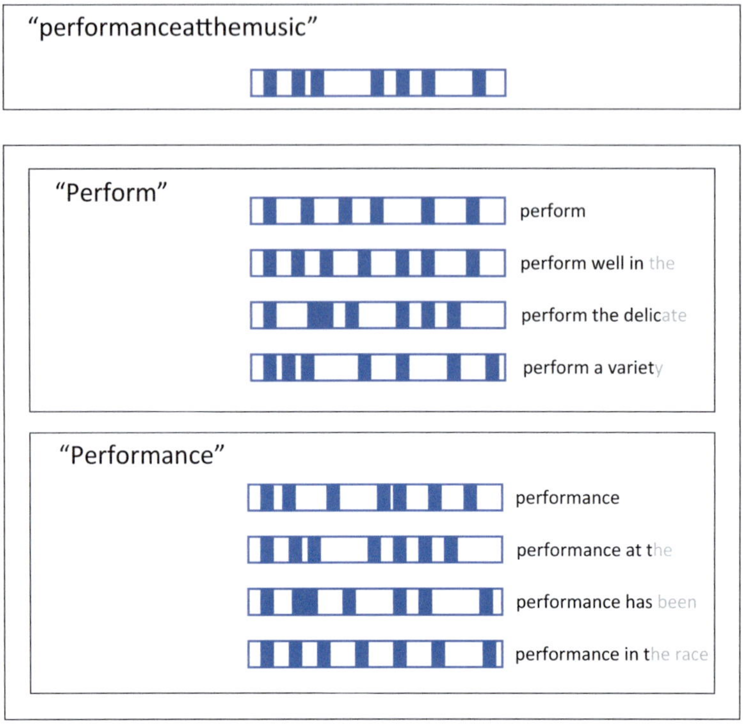

Fig. 7.2 Example for surround text matching

7.3.3 Handling Unknown Words

In order to manage unknown words, both the THDICTSDR and THSDR algorithms employ the observation that these words are generally short and infrequent. Furthermore, the presence of an unknown word often suggests previous word segmentations may be flawed, necessitating backtracking. Hence, when an unknown word is discovered, the algorithms designate two anchor words on either side of the unknown word based on their length and frequency. For instance, if the recognized words and their frequencies are ["I", 10], ["LO", 3], and ["BATS", 9], and the text is "ILOVEBATS", the word segmentation is "I", "LO", "VE", "BATS", with "VE" being the unknown word. The algorithm sets "I" and "BATS" as anchor words due to their high frequency or lengthy nature. The algorithm then searches for neighboring words and finds that "LO" is infrequent, thus "LO" and "VE" are combined to form the word, "LOVE" (Fig. 7.3).

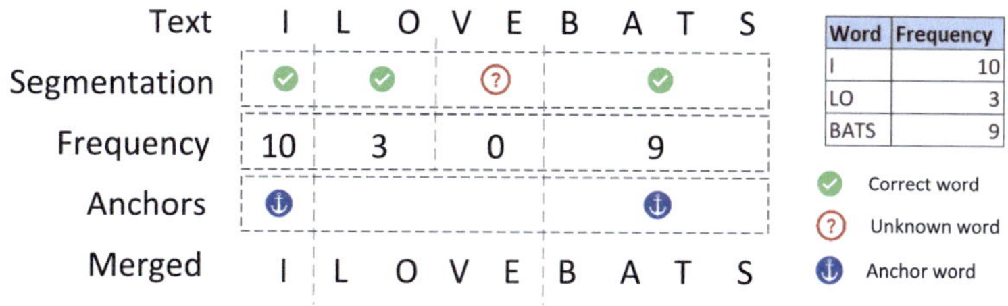

Fig. 7.3 Handling unknown words by setting anchors and merging

7.3.4 N-Gram

In an effort to enhance accuracy performance, the author also incorporated n-grams. However, they noticed that there were inconsistencies in word segmentations within the training data. This is a recognized issue in the field, wherein even a single person may segment a word differently, resulting in data inconsistencies. To mitigate this, the algorithm examines the co-occurrence value of the n-grams. For instance, if the frequency of "icecream" is higher than the co-occurrence value of "ice cream", then the two words "ice cream" will be consolidated into "icecream". For the purpose of this chapter, specifically 2-g were utilized to choose the word.

7.4 Evaluation

The effectiveness of the newly proposed THDICTSDR and THSDR methods is evaluated through comparison with established techniques, including dictionary-based methods, longest matching, newmm, and state-of-the-art approaches such as Deepcut.. By assessing and contrasting these methods, the study aims to provide insights into the performance and advantages of THDICTSDR and THSDR in Thai word segmentation. We opted to utilize Deepcut, given its acknowledgment as state-of-the-art and its superior accuracy. As an alternative, we also employed Attacut. However, as alluded to in the paper, Attacut does not outperform Deepcut in terms of precision. Moreover, Deepcut is a user-friendly, open-source tool in ThaiNLP. The evaluation process is executed using two datasets, Best2010 [17] and LST20 [18], on an ASUS TUF A15 laptop equipped with an AMD Ryzen 7 5800H, 8 CPU cores, 16 threads, 32GB of RAM, and a GPU RTX3060 6GB. The Best2010 dataset is composed of 415 Thai documents, roughly 5.1 million words, and a 104k vocabulary, spanning four domains such as articles, news, encyclopedias, and novels. Conversely, the LST20 Corpus provides five layers of linguistic annotation, comprising word boundaries, POS tagging, named entities, clause boundaries, and sentence boundaries. It contains 3,164,002 words, 288,020 named entities, 248,181 clauses, and 74,180 sentences. The evaluation is conducted independently for each dataset, with 90% of the data utilized for training and the remaining 10% for testing. The effectiveness of the

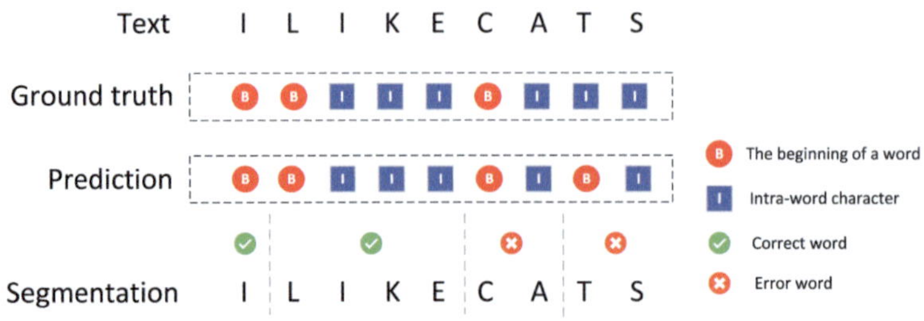

Fig. 7.4 Word-Level calculation of evaluation metrics

proposed methods is compared with existing techniques based on criteria such as accuracy, precision, recall, and F1-score.

Deepcut was retrained using 90% of the training dataset to eliminate the possibility of including testing data from its pre-trained version. Similarly, THSDR relies on the Lexitron Thai-English Dictionary and further learns from the Best2010 training data to generate SDRs.

In order to evaluate the new model, five metrics are employed at the word level. These metrics are Precision, Recall, F1-Score, Intersection over Union (IoU), and Processing time. Precision, Recall, F1-Score, and IoU are computed using the example illustrated in Fig. 7.4 and the Eqs. 7.1, 7.2, 7.3, and 7.4.

During the evaluation process, two parameters were examined: the span of the surrounding context and the magnitude of the SDR. For the length of the surrounding context, three values were tested: 8, 16, and 32. Similarly, for the size of the SDR, three values were assessed: 1024, 2048, and 4096. The outcomes revealed that the highest performance was achieved with a surrounding context length of 16 and an SDR size of 2048. Consequently, these values were chosen for the final experiment.

$$\text{Precision} = \frac{\text{the number of correct words}}{\text{the number of word predictions}} \tag{7.1}$$

$$\text{Recall} = \frac{\text{the number of correct words}}{\text{the number of words in the ground truth}} \tag{7.2}$$

$$\text{F1-Score} = \frac{2 \times \text{Recall} \times \text{Precision}}{\text{Recall} + \text{Precision}} \tag{7.3}$$

$$\text{IoU} = \frac{TP}{TP + FP + FN} \tag{7.4}$$

The equation showcased earlier clarifies that TP (True Positive) corresponds to the count of word segments accurately identified, whereas FP (False Positive) corresponds to the count of word segments that were wrongly identified. FN (False Negative), on the other hand, symbolizes the count of word segments that the method failed to identify.

The performance assessment of the first technique, which is a dictionary-based method, involved three distinct test conditions. The first condition saw the method learning solely from the training dataset and a common lexicon. The second condition presumed that the method had been exposed to all the words in the sentences used for user-interactive applications, or it had learned a vocabulary extensive enough to encompass them. The third condition entailed that the method was aware of all word connections and sentences. The second method was tested to showcase its resilience to noise. Noise, generated from the Best dataset at four distinct levels: 1, 3, 5, and 10%, was introduced and the method's performance was measured in these conditions.

7.4.1 THDICTSDR

Table 7.1 presents the results of the first evaluation, clearly highlighting that the Precision, Recall, Intersection over Union (IoU), and F1-Score on the Best dataset for the Longest Matching and Newmm methods are significantly lower compared to Deepcut and THDICTSDR. Although THDICTSDR demonstrates a higher recall performance of 96.89% compared to Deepcut, its precision, IoU, and F1-Score are slightly lower. Furthermore, the processing time of THDICTSDR is longer than that of Deepcut due to the extensive number of rules it needs to check. Nevertheless, these results indicate that THDICTSDR is comparable to state-of-the-art methods and offers a unique alternative. However, it should be noted that the precision of THDICTSDR may be affected by the problem of unknown words. Despite efforts to address this issue via Handling Unknown Words method, some incorrect segmentations may still occur, resulting in a slight decrease in precision.

Table 7.2 demonstrates the outcomes from the second evaluation, where THDICTSDR was assessed using the LST20 dataset. These results align with the findings from the initial evaluation. Although the recall and precision of THDICTSDR were slightly lower than Deepcut, it still demonstrated high recall. Nevertheless, THDICTSDR showed a better performance when compared to the Longest and Newmm methods, further validating the effectiveness of the THDICTSDR approach.

Table 7.3 shows the results of the third evaluation, assessing the effectiveness of THDICTSDR's vocabulary knowledge. The significance of extensive vocabulary knowledge for precise word segmentation is emphasized, mirroring human

Table 7.1 Word-Level performance comparison on Best2010

Method	Train data	Test data	Precision (%)	Recall (%)	F1-Score (%)	IoU	Time (s/word)
Longest	Train	Test	77.01	67.05	71.68	0.56	3.95E-05
newmm	Train	Test	80.30	69.96	74.77	0.60	7.34E-06
Deepcut	Train	Test	96.11	96.57	96.34	0.93	1.50E-03
THDICTSDR	Train	Test	94.32	96.89	95.60	0.92	0.03

Table 7.2 Word-Level performance comparison on LST20

Method	Train data	Test data (%)	Precision (%)	Recall (%)	F1-Score	IoU	Time (s/word)
Longest	Train	Test	80.60	72.06	76.09	0.61	4.20E-05
newmm	Train	Test	83.28	73.76	74.77	0.64	7.45E-06
Deepcut	Train	Test	96.67	97.13	96.90	0.94	1.50E-03
THDICTSDR	Train	Test	94.42	96.85	95.62	0.92	0.028

Table 7.3 Word segmentation performance on all learned vocabularies

Method	Train data	Test data (%)	Precision (%)	Recall (%)	F1-Score	IoU	Time (s/word)
THDICTSDR	Best train	Best test	96.08	97.50	96.78	0.94	0.03
THDICTSDR	LST train	LST test	95.48	97.50	96.40	0.93	0.03

proficiency in language understanding. In cases where an unknown vocabulary word appears during segmentation, the software has been designed to ask the user to confirm the new word. This method is distinct from the deep learning model's approach of labeling segmentation words to learn error patterns.

As per the evaluation results, THDICTSDR surpassed Deepcut in the Recall, IoU, and F1-Score metrics, reaching 97.50%, 0.94, and 96.78% respectively on the Best dataset, and 97.50% on Recall for the LST dataset. Despite these impressive results, THDICTSDR's performance on the LST dataset in terms of precision, IoU, and F1-Score was slightly lower. This is attributed to the approach THDICTSDR employs to manage unknown words, indicating an area of possible improvement. Nevertheless, these findings validate the effectiveness of the vocabulary-learning approach used by THDICTSDR.

Table 7.4 demonstrates the effectiveness of THDICTSDR when the algorithm has been trained on the correct words and word connections in the training data, mirroring a human who has fully grasped all words and connections. In this specific scenario, THDICTSDR exhibits a notable improvement in its performance, achieving an impressive F1-score of 99.48% on the Best dataset and 99.37% on the LST dataset. When compared to Deepcut, which achieved an F1-score of 97.65%, it's clear that the improved learning approach of THDICTSDR provides a considerable advantage in terms of performance.

Table 7.4 Word segmentation performance on all learned vocabularies

Method	Train data	Test data (%)	Precision (%)	Recall (%)	F1-Score	IoU	Time (s/word)
Deepcut	Best train	Best train	97.42	97.89	97.65	0.95	1.50E-03
THDICTSDR	Best train	Best train	99.20	99.55	99.48	0.99	0.03
THDICTSDR	LST train	LST train	98.36	99.69	99.37	0.98	0.03

7.4.2 THSDR

Table 7.5 illustrates the findings of THSDR's follow-up experiment to Table 7.4. In this experiment, noise is introduced into the data through character replacement, insertion, deletion, and transposition, at varying percentages of 1%, 3%, 5%, and 10%. Remarkably, the results highlight the exceptional fault tolerance of THSDR compared to Deepcut, a deep learning approach, across all noise scenarios. THSDR demonstrates superior performance and resilience in handling noise, reinforcing its effectiveness as a robust solution for Thai word segmentation.

7.4.3 Parameters

In Table 7.6, the evaluation of different SDR sizes (1024, 2048, and 4096) reveals that there is no significant difference in performance or processing time among these sizes. Changing the SDR size has minimal impact on overall performance. This suggests that increasing the SDR size to enhance the capacity for the union operation in merging could potentially improve performance without sacrificing processing time. However, in this study, the best F1-Score was achieved with an SDR size of 2048.

On the other hand, Table 7.7 presents the results obtained by varying the word length threshold to 64, 32, 16, and 8. The findings demonstrate that increasing the word length can have a negative impact on performance, as it leads to encoding too many words into the SDR, resulting in multiple matches. Conversely, reducing the word length can also adversely affect performance, as it results in fewer surrounding

Table 7.5 Word segmentation performance on all learned vocabularies

Method	Train data	Test data (%)	Precision (%)	Recall (%)	F1-Score	IoU	Time (s/word)
Deepcut	Best	Original	97.7	98.12	97.91	0.96	0.002
THSDR	Best	Original	99.76	99.84	99.8	0.99	0.04
Deepcut	Best	Noise 1%	92.99	93.50	93.25	0.87	0.002
THSDR	Best	Noise 1%	97.89	98.00	97.95	0.96	0.04
Deepcut	Best	Noise 3%	88.77	89.22	88.99	0.80	0.002
THSDR	Best	Noise 3%	93.44	93.70	93.57	0.88	0.04
Deepcut	Best	Noise 5%	84.62	85.00	84.81	0.74	0.002
THSDR	Best	Noise 5%	90.50	91.02	90.76	0.83	0.04
Deepcut	Best	Noise 10%	74.96	75.03	74.99	0.60	0.002
THSDR	Best	Noise 10%	79.29	82.00	80.62	0.68	0.04

Table 7.6 Word segmentation performance for varying SDR sizes

SDR size	Word length	Precision (%)	Recall (%)	F1-Score (%)	IoU	Time (s/word)
1024	16	93.47	96.82	95.12	0.91	0.03
2048	16	94.32	96.91	95.60	0.92	0.03
4096	16	93.64	96.82	95.20	0.91	0.03

Table 7.7 Word segmentation performance for varying word length

SDR size	Word length	Precision (%)	Recall (%)	F1-Score (%)	IoU	Time (s/word)
2048	64	83.36	93.53	88.15	0.79	0.03
2048	32	93.05	96.54	94.76	0.90	0.03
2048	16	94.32	96.91	95.60	0.92	0.03
2048	8	81.31	75.39	78.24	0.64	0.03

context words. Based on these results, a word length threshold of 16 demonstrates the best performance and has been chosen for use in the experiment.

7.4.4 Complexity and Time

Complexity and time in the context of THSDR can be divided into two components: training time and prediction time. During training, the training data is encoded into SDRs and stored in memory. As a result, the complexity of training time is denoted as O(nh), where n represents the number of training data points and h represents the number of hash functions utilized.

On the other hand, the complexity of prediction time is expressed as O(ns), where n signifies the number of training data points and s represents the size of the SDR. This indicates that prediction time scales linearly with the number of training data points and the size of the SDR.

To mitigate processing time and enhance efficiency, techniques such as union operations and parallel processing can be implemented. These approaches are demonstrated in Chap. 6, and can contribute to reducing the overall processing time of the THSDR algorithm.

7.5 Conclusion

This chapter introduces two novel methods that utilize Sparse Distributed Representations (SDRs) to simulate the learning process of the brain. These methods demonstrate higher accuracy compared to traditional dictionary-based approaches in Thai word segmentation. The first method achieves results that are comparable to the state-of-the-art method, Deepcut, and in certain cases, even outperforms it. However, it should be noted that the processing time of this method is longer than other approaches. There is room for improvement in enhancing the computational efficiency of this method to better support real-time or large-scale applications.

The second method exhibits good resilience to noise, making it well-suited for applications such as Thai Optical Character Recognition (OCR). It can effectively handle and accurately segment words even in the presence of noise, showcasing its robust performance.

The combination of dictionary-based and learning-based methods in these approaches provides accuracy that is on par with, and in some cases better than, state-of-the-art techniques like Deepcut. The advantage of the dictionary-based approach lies in its ability to quickly learn new vocabulary. However, state-of-the-art methods require explicit word boundaries to learn a new error model from training data. Overall, these novel methods offer advancements in Thai word segmentation, delivering higher accuracy and resilience to noise. Further improvement can focus on reducing processing time, refining model efficiency, and enhancing the adaptability of the combined approach.

Building upon the exploration of spelling correction and Thai word segmentation using the brain-inspired model, the final chapter transitions from task-specific evaluation to a broader reflection on the overall contributions of this book. After presenting neuroscience-inspired concepts, proposing a simplified learning algorithm, and validating it through practical use cases, , Chap. 8 offers a concluding discussion on how these ideas could contribute to addressing some of the challenges faced by current AI approaches. While the current results are encouraging, many open challenges remain, along with opportunities for further refinement and integration with existing AI methods.

References

1. Tapsai, C., Unger, H., Meesad, P.: Thai natural language processing—word segmentation, semantic analysis, and application. In: Studies in Computational Intelligence, vol. 918, pp. 1–22. Springer, Cham (2021). https://doi.org/10.1007/978-3-030-56235-9
2. Pornprasertsakul, A.: Thai Syntactic Analysis. Ph.D. thesis, Asian Institute of Technology, Thailand (1994)
3. Paripremkul, K., Sornil, O.: Segmenting words in Thai Language using minimum text units and conditional random field. J. Adv. Inf. Technol. **12**(2), 135–141 (2021). https://doi.org/10.12720/jait.12.2.135-141

4. Kittinaradorn, R., Chaovavanich, K., Achakulvisut, T., Srithaworn, K., Chormai, P., Kaewkasi, C., Ruangrong, T., Oparad, K.: DeepCut: a Thai word tokenization library using Deep Neural Network. Zenodo (2019). https://doi.org/10.5281/zenodo.3457707
5. Chormai, P., Prasertsom, P., Rutherford, A.T.: AttaCut: A Fast and Accurate Neural Thai Word Segmenter (2019). arXiv:1911.07056style
6. Thairatananond, Y.: Towards the Design of a Thai Text Syllable Analyzer. Master's thesis, Asian Institute of Technology, Thailand (1981)
7. Chanyapornpong, S.: A Thai Syllable Separation Algorithm. Master's thesis, Asian Institute of Technology, Thailand (1983)
8. Poovarawan, Y., Imarrom, W.: Thai syllable separator by dictionary. In: Proceedings of the 9th Annual Meeting on Electrical Engineering of the Thai Universities, pp. 1–6, Khon Kaen, Thailand (1986)
9. Sornlertlamvanich, V.: Word segmentation for Thai in machine translation system. In: Proceedings of the International Conference on Machine Translation, pp. 556–561 (1993)
10. Raruenrom, S.: Dictionary-based Thai Word Separation. Master's thesis, Chulalongkorn University, Thailand (1991)
11. Kawtrakul, A., et al.: A Lexibase model for writing production assistant system. In: Proceedings of the 2nd Symposium on Natural Language Processing, pp. 226–236, Bangkok, Thailand (1995)
12. Meknavin, S.: Towards 99.99% accuracy of Thai word segmentation. Oral presentation at the Symposium on Natural Language Processing in Thailand, Bangkok, Thailand (1995)
13. Haruechaiyasak, C., Kongyoung, S., Dailey, M.: A comparative study on Thai word segmentation approaches. In: 2008 5th International Conference on Electrical Engineering/Electronics, Computer, Telecommunications and Information Technology (ECTI-CON), vol. 1, pp. 125–128. IEEE (2008). https://doi.org/10.1109/ECTICON.2008.4600388
14. Aroonmanakun, W.: Collocation and Thai word segmentation. In: Proceedings of SNLP and Oriental COCOSDA Workshop, pp. 68–75 (2002)
15. Theeramunkong, T., Sornlertlamvanich, V., Tanhermhong, T., Chinnan, W.: Character cluster based Thai information retrieval. In: Proceedings of the Fifth International Workshop on Information Retrieval with Asian Languages, pp. 75–80 (2004). https://doi.org/10.1145/355214.355225
16. Boonkwan, P., Supnithi, T.: Bidirectional deep learning of context representation for joint word segmentation and POS tagging. In: Nguyen, N.T., Janiak, A., Chbeir, R. (eds.) Advanced Computational Methods for Knowledge Engineering, pp. 184–196. Springer, Cham (2018)
17. Thailand National Electronics and Computer Technology Center (NECTEC): Benchmark for Enhancing the Standard of Thai Language Processing 2010, BEST 2010 (2010). https://aiforthai.in.th/corpus.php. Accessed 4 May 2025
18. Boonkwan, P., Luantangsrisuk, V., Phaholphinyo, S., Kriengket, K., Leenoi, D., Phrombut, C., Boriboon, M., Kosawat, K., Supnithi, T.: The Annotation Guideline of LST20 Corpus (2020). arXiv:2008.05055

Chapter 8
Conclusion and Future Work

This chapter concludes the book by evaluating the broader implications of the proposed brain-inspired model. Section 8.1 summarizes key findings from previous applications and highlights the model's strengths in efficiency, noise tolerance, and adaptability. The chapter introduces several insights, including the benefits of binary computation, context-based representation, and scalable learning mechanisms. Section 8.2 outlines future directions, such as hardware implementation, improved online learning, and integration with neuroscience principles. These reflections highlight the potential of brain-inspired models to help address some limitations of current AI systems, contribute to the refinement of existing methods, and inspire the development of novel approaches.

8.1 Conclusion

Despite rapid progress, current AI systems still face significant limitations, such as high energy consumption, reliance on large datasets, low tolerance to noise, and limited adaptability and transparency. These challenges suggest that AI continues to fall short of human intelligence in several important dimensions. This book seeks to address these issues by drawing deeper insights from the human brain, which remains the most efficient and adaptable learning system known. These limitations serve as a catalyst to re-examine the brain's architecture and inspire improvements to existing AI methods, or even the development of entirely new approaches. Rather than reinventing the wheel, the book builds upon Hawkins' foundational ideas and adapts them for sequential learning tasks in natural language processing. The proposed method emphasizes simplicity, emulating the brain's mechanisms while delivering practical and reliable results.

T. Soisoonthorn and H. Unger, *A Brain-Inspired Approach to Natural Language Processing*, Studies in Computational Intelligence 1225,
https://doi.org/10.1007/978-3-032-00014-9_8

This book has taken readers on a progressive journey, from the biological foundations of intelligence to practical applications of brain-inspired computational models. Chapter 1 laid the groundwork by exploring how the human brain achieves intelligence through efficiency, adaptability, and hierarchical memory processing. Chapter 2 deepened this understanding by examining the neocortex and its role in learning and prediction, culminating in Jeff Hawkins' theoretical framework that links brain structure to computational principles. Chapter 3 then provided a critical overview of traditional and modern AI methods, identifying their limitations and motivating the need for models that better emulate biological intelligence. This led to Chap. 4's introduction of Sparse Distributed Representations (SDRs), a biologically plausible encoding mechanism that underpins robust and efficient memory. Chapter 5 extended these ideas into the temporal domain, proposing a novel sequence learning model inspired by neural processes such as working memory and forgetting, and demonstrating its practicality through hardware simulation. Chapters 6 and 7 validated the proposed model through real-world tasks, including spelling correction and Thai word segmentation, revealing not only improved accuracy but also greater noise tolerance and efficiency on limited-resource devices.

In addition, this book integrates key insights from both neuroscience and artificial intelligence into a unified narrative, enabling readers to develop a deeper understanding of how biologically inspired principles can drive the creation of more efficient and adaptable AI systems. Section 3.9 presents a comprehensive review of current AI approaches, analyzing their strengths and limitations while offering guidance on how neuroscience concepts can be applied to overcome existing challenges. The aim is to provide a meaningful shortcut, equipping readers with foundational knowledge and practical direction so they can build upon established work rather than starting entirely from scratch.

The results of the proposed brain-inspired algorithm are presented in Chaps. 6 and 7, where it demonstrates performance that is comparable to, and in some cases better than, state-of-the-art methods. These chapters showcase the algorithm's effectiveness across different natural language processing tasks, validating its adaptability and robustness. The observed successes and limitations of the approach are closely tied to several key findings, which offer deeper insight into both its strengths and the areas that warrant further refinement.

- **Bits Operate Faster Than Floating-Point Arithmetic**
  One major contributor to the algorithm's efficiency is its reliance on binary operations. Since CPUs and digital circuits are optimized for binary computation, bitwise operations are executed faster and require less memory compared to floating-point arithmetic. Representations using only "0" and "1" resemble how neurons in the brain fire in binary-like spikes. While this approach emphasizes binary processing, it does not imply that weights are irrelevant to all brain mechanisms. Rather, for this model, binary representations offer practical and performance benefits.
- **Connections Between Neurons Form Representations**
  Traditional encoding methods focus on transforming input signals into representations. In contrast, this algorithm also treats the connections between sequences

as meaningful representations. This mirrors how humans understand relationships between events, allowing the system to capture contextual meaning and improve noise tolerance. This novel perspective, particularly in the context of NLP, extends beyond signal encoding and is rarely used outside of commercial or numeric applications. It introduces a fresh take on how memory and association may work biologically and computationally.

- **Scalability to Big Data**
  Handling large-scale data presents challenges for any algorithm. To address this, two mechanisms are introduced: union (merging) and parallel processing. The union method, which combines multiple patterns into a single representation, can slightly impact precision but improves scalability. Parallel processing, on the other hand, increases throughput without degrading performance. These mechanisms reflect how the brain manages massive information streams and offer a promising direction for handling Big Data. Future improvements could focus on incorporating dimensionality reduction techniques to enhance processing efficiency while preserving accuracy.
- **Learning and Error Model**
  Chapter 6 highlights a fundamental limitation in conventional deep learning: its dependence on error-driven learning and repeated exposure to labeled data. In contrast, humans often learn from minimal examples and generalize effectively. The proposed algorithm reflects this by enabling it to capture knowledge with minimal training and without relying on explicit error labels. This human-like efficiency marks a significant departure from traditional supervised learning paradigms.
- **Decision-Making in Online Learning**
  Effective online learning requires mechanisms to determine when information should transition from short-term to long-term memory. This model adopts a threshold-based approach, where learning is solidified based on the significance or repetition of input. This mimics human behavior, such as remembering commonly searched phrases. However, thresholds alone may not always distinguish between correct and incorrect inputs. To address this, future research could incorporate a reward-based mechanism, enabling more adaptive and context-aware learning decisions.

8.2 Future Work

While the proposed algorithm has demonstrated good performance on the tasks presented, it is still far from complete. The remaining challenges highlight opportunities for future work.

- **Hardware Implementation** This book presents only a hardware simulation. Future work should focus on actual hardware implementation, which is a key step toward supporting real-time applications and validating the model's efficiency in practical environments.

- **Enhancing Online Learning** In its current form, the algorithm's online learning relies mainly on frequency to determine the correctness of inputs. However, this mechanism may be limited in certain scenarios. Inspired by neuroscience, incorporating a reward-based reinforcement mechanism could offer a more dynamic and context-aware approach to learning.
- **Efficient SDR Encoding for NLP** The current SDR encoding method for NLP is based on hashing characters and their relationships. To reduce memory usage and processing time, future work could explore syllable-based patterns instead of character-level encoding. This strategy could also be extended to other types of sequence data by identifying domain-specific patterns.
- **Thai Spelling Correction and Word Segmentation** The existing Thai word segmentation model works effectively with correctly spelled words but fails to handle misspellings, treating them as unknown tokens. This limitation presents challenges in Thai OCR systems, which often require both spelling correction and segmentation. A combined approach, drawing from the techniques in Chaps. 6 and 7, could improve robustness in handling both tasks simultaneously.
- **Integrating AI Approaches with Neuroscience Principles** As discussed in Sect. 3.9, there are several identified gaps in current AI methodologies. Combining traditional AI techniques with neuroscience-inspired principles could lead to improvements in areas such as self-supervised learning, noise tolerance, adaptive architectures, neuron-level independence, and explainable AI.

MIX
Papier aus verantwortungsvollen Quellen
Paper from responsible sources
FSC® C105338
FSC
www.fsc.org

If you have any concerns about our products,
you can contact us on
ProductSafety@springernature.com

In case Publisher is established outside the EU,
the EU authorized representative is:
Springer Nature Customer Service Center GmbH
Europaplatz 3, 69115 Heidelberg, Germany

Printed by Libri Plureos GmbH
in Hamburg, Germany